Mobil
BERYL A

ENERGY

Marshall Cavendish London & New York

Edited by Donald Clarke
Published by Marshall Cavendish Books Limited
58 Old Compton Street, London W1V 5PA

Printed in Great Britain

ISBN 0 85685 432 8

This material has previously appeared in the publication *How It Works*

INTRODUCTION

The increasing scarcity and higher cost of fossil fuels today has lent urgency to the arguments of conservationists. The fact is simply that we shall run out of oil someday, and even coal reserves are limited. What will we do then?

What is energy? Where do we get it from and how do we use it? As this is written, there is news of a proposed tide mill on the River Severn, which has the highest tides in Europe; this would be the biggest engineering project in history. The National Engineering Laboratory is reportedly building an experimental wave power station, which would use 'nodding duck' devices to extract the energy from the waves. How do these things work? Are they the answer to the energy problem?

The truth is that energy fundamentally resides in the structure of matter. What we call electricity is the motion of clouds of electrons, which are constituent parts of all atoms. But they do not flow by themselves. The answer to the energy problem almost certainly resides in new and more efficient methods of extracting it. It is a little like a detective story: will the new techniques be developed before the industrialized countries suffer a dramatic fall in their standard of living?

The whole story, from the dramatic discoveries of 19th century scientists to the details of today's electricity generation, from the mysteries of the Theory of Relativity to solar power and particle research, is told in this book, *Energy*.

CONTENTS

DEFINITION AND THEORY

ENERGY

Energy is *work* in its broadest scientific sense. A bullet in motion possesses energy by virtue of its motion and this *kinetic* energy is 'given up' or transferred on hitting a target. The energy of the bullet goes into deforming or breaking the target, that is, doing work on the target, and as heat and sound. A bullet at rest does not possess this energy.

A mass raised against gravitational force possesses *potential* energy because it has a 'potential' to do work, that is, its potential energy can be used to do something, such as hammer a pile into the ground.

Kinetic and potential energy are both forms of stored energy relating to the motion of bodies or their potential to move, and as such form the basis of mechanical physics. There are, however, many other forms in which energy exists. A drum of oil is inert when left to itself, yet it contains latent (undeveloped) chemical energy which can be used in a diesel engine, for example, to drive a train. The gunpowder behind the bullet possesses latent energy before it is fired. Energy is stored in the magnetic field of a permanent magnet because it will move a piece of iron in the vicinity—thus doing work on the iron. Also, energy can be transferred from place to place in the form of radiation. This can be heat radiation, or light, which is a form of electromagnetic radiation. It is in this form that the Earth receives energy from the Sun.

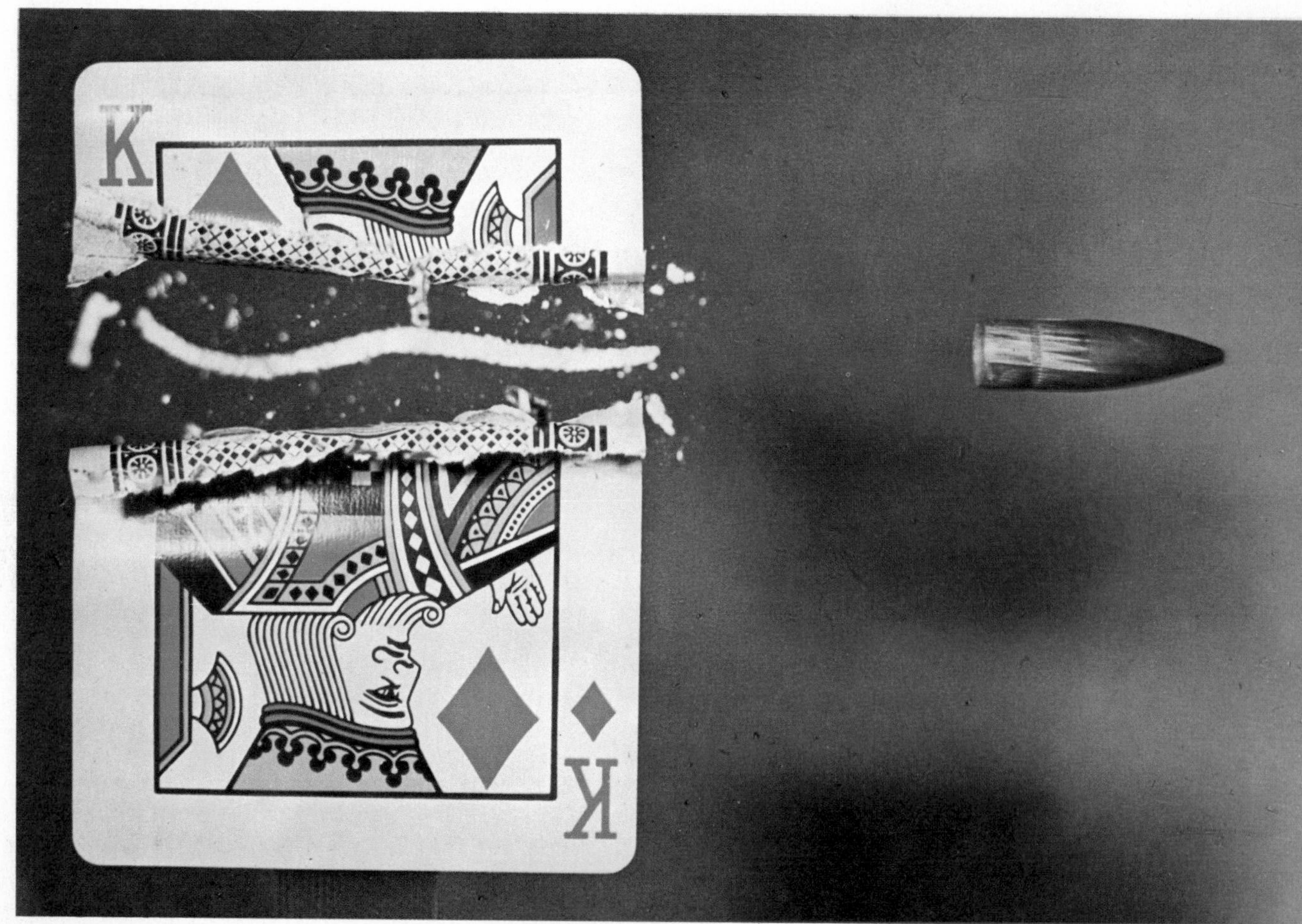

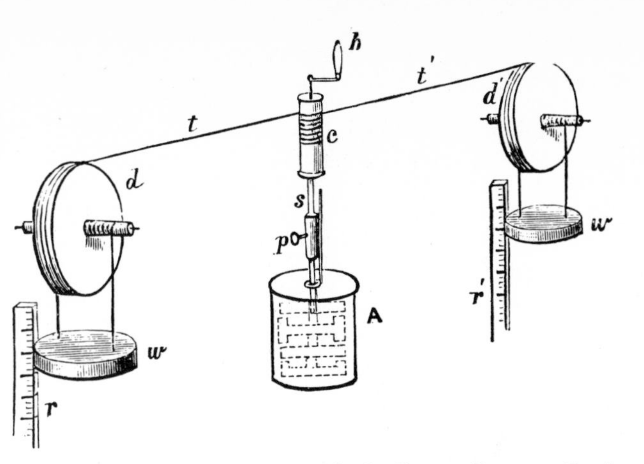

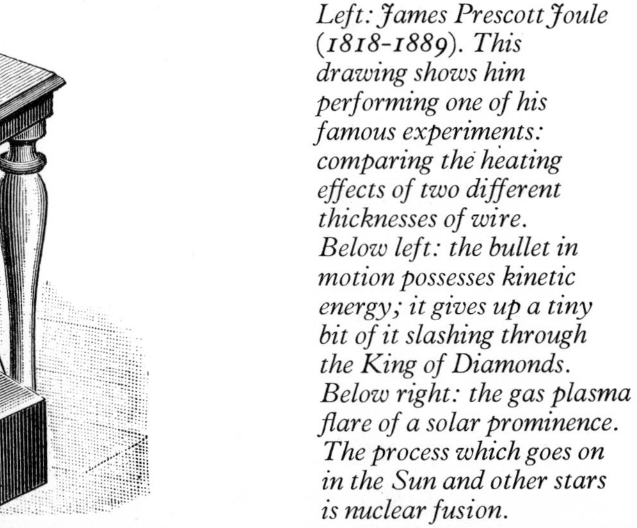

Left: James Prescott Joule (1818-1889). This drawing shows him performing one of his famous experiments: comparing the heating effects of two different thicknesses of wire.
Below left: the bullet in motion possesses kinetic energy; it gives up a tiny bit of it slashing through the King of Diamonds.
Below right: the gas plasma flare of a solar prominence. The process which goes on in the Sun and other stars is nuclear fusion.

Energy is convertible from one form to another—in some cases with ease, as in the pendulum which interchanges potential and kinetic energy during its cyclical swinging. In other situations, man-made energy conversion devices are required. In an electrical generating station, for example, the chemical energy of coal or oil is released by combustion as heat to raise steam, converted into rotary energy in a turbine, then again converted into electrical energy in an electromagnetic generator.

Such conversions are never fully effective. For example, in a power station only about 40% of the latent energy of the fuel is converted into electricity. But if the energy 'lost'—which eventually becomes low grade heat—is accounted for in the energy balance-sheet, the total quantity of energy sums to the same amount on both sides. This is a statement of the principle of energy conservation.

The nature of energy

It is impossible to say what energy is because energy, like time, is a concept so basic that there are no terms available more fundamental to describe it. It can only be quantified in more basic units, for example, by relating it to mass, velocity, temperature and so on. Yet although the concept is the cornerstone of modern scientific thinking, it is little more than a century old.

Sir Isaac Newton, in formulating his epoch-making laws

Right: mass is an unchanging property, but weight depends on the environment. Going up and down in a lift, one's weight changes, but mass stays the same.

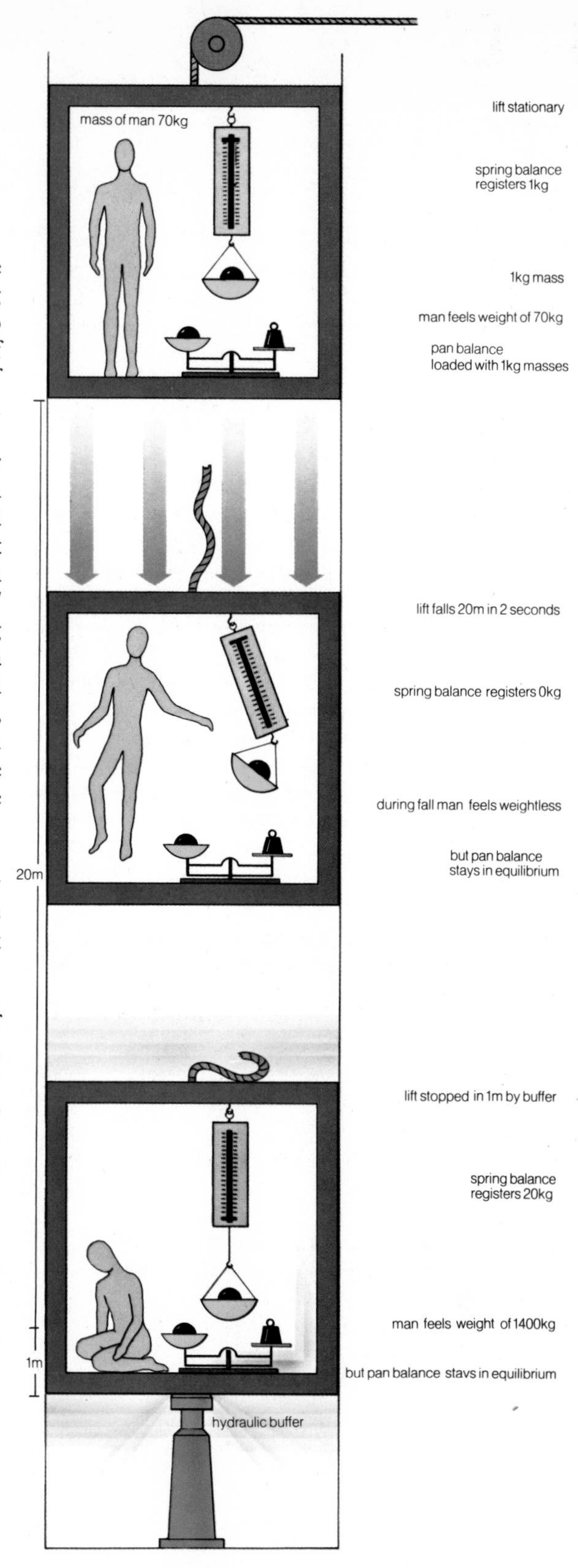

of motion, did not mention energy. The term (from the Greek word meaning work) was coined by Thomas Young (1773—1829) eighty years after Newton and applied to what is now called the kinetic energy of a body. A body of mass m moving with velocity v has a kinetic energy of $\frac{1}{2}mv^2$. Half a century later, Rankine coined the term potential energy.

Both these terms concern mechanical physics, and their relation to thermal energy (heat) was not realized until Joule demonstrated two crucial experiments. He showed that the heat produced by the passage of an electric current through a wire was related to the square of the current and also that heat was produced by mechanical work. In 1847 he obtained the mechanical equivalent of heat by measuring the temperature rise in water resulting from the action of a paddle driven by a falling mass. This experiment convinced Lord Kelvin, although Carnot, father of the heat engine, never accepted that heat was other than a 'fluid', called caloric, possessed by 'hot' bodies. After Joule, the principle of energy convertibility gained rapid ground and with it the concept of energy conservation.

Energy and mass

Until Einstein announced his theory of relativity, the energy conservation concept remained unassailed. Along with this grew the idea of the conservation of mass which states that matter can be rearranged but not destroyed.

Einstein's theory modified both of these fundamental concepts. He showed that every physical occurrence, of whatever kind, can be specified completely only if it is known when as well as where it occurred. A physical 'event' is placed not only in the three dimensions of space but also in the fourth dimension of time. A body in motion therefore exists in such a system and its velocity relative to a (stationary) observer is important in determining the properties of that body. There is a limiting factor here, however, as nobody can travel faster than the speed of light (usually denoted c) and as a body approaches this speed both its observed mass and energy tend to become infinitely large. This led Einstein to the conclusion that a mass m at rest is equivalent to an amount of energy given by $m \times c^2$, but in motion the effective mass (and therefore the effective energy) increases according to the velocity v of the body in relation to the velocity of light c.

The equivalence of mass and energy has been triumphantly verified. It elucidates the phenomena of radioactivity, the explosion of the atomic bomb and the production of energy by nuclear fission. Where such a reaction takes place an enormous amount of energy is released accompanied by an actual reduction in total mass. Thus the production of vast quantities of radiant energy emitted by the stars is the result of a conversion of a small fraction of their mass into radiation.

Units of energy

The historic growth, side by side, of branches of physics, between which any interrelation remained unsuspected, led to many different scientific and legal units in which energy can be specified. Examples are calories, therms and British thermal units (for heat), watt-hours (for electrical energy), foot-pounds and kilogram-metres (for mechanical energy), gauss-oersteds (for the permanent magnet industry) and several others. The modern view of the unit of energy, together with the awkward numerical conversion factors otherwise necessary, led to the adoption of a single basic unit. This is the joule, named after the scientific amateur from Manchester who demonstrated energy equivalence in the middle of the 19th century.

POWER

Power is defined as the *rate of doing work* and is important in assessing the 'value' of a machine or device. When used in a scientific context, the concept of power does not differ very much from that met in everyday language. A 'powerful' man can be expected to finish a particular manual job more quickly than a less well-built man; a 'powerful' computer can compute more steps in a second than its predecessor.

Force, work and power

Consider the example of a battery-driven motor used for raising loads up a cliff. The first consideration is the maximum load that the motor can lift. Load is *force* and is measured in such units as newtons (N) or pounds-force (lbf). Secondly, it is important to know how much work the motor can do before the battery is run down (discharged). The work done is equal to the lifting force multiplied by the distance moved. Work is the same as energy and is measured in such units as newton-metres (Nm) or foot-poundsforce (ft lbf). The internationally agreed unit of energy is the *joule* which is defined as one newton-metre.

For example, if the load to be raised is 100 N in weight (equivalent to 25 lbf) and the battery runs out after the load has been lifted 100 metres, then the work done is 10,000 Nm (or 7500 ft lbf) which is 10,000 J. This work done by the motor is the same as the energy gained by the load—in this case potential energy. If the motor is 100% efficient in its conversion of electrical energy to mechanical energy (this is an ideal state) then the total stored energy of the battery before lifting the load is 10,000 J.

So far, no consideration has been given to the speed at which the load was lifted. The third important factor of a machine is therefore how fast the machine can work—this factor is the *power* of the machine.

The basic definition of power is:

$$\text{power} = \text{work done} \div \text{time taken}$$

but work done is equal to force exerted × distance moved so:

$$\text{power} = \text{force exerted} \times \text{speed of movement.}$$

Any one of these formulas can be used for calculating power. If the load in the above example is raised in 5 minutes (300 seconds) then the power developed by the motor is about 33 joules per second (33 Js^{-1}). It can be seen that the faster the load is lifted the more power required, although the total work done (at any speed) remains the same.

Electrical power

Electrical power is generally expressed in watts, where one watt (1 W) is equal to one joule per second. In the case of an electrical component such as a resistor, the power consumed, measured in watts, is the product of the voltage across the component and the current flowing through it. The voltage (V) across a resistor is equal to the current (I) multiplied by the resistance (R). Thus, the power consumed by a resistor (this power is dissipated in the form of heat) is I^2R or V^2/R. A 58 ohm resistor will therefore dissipate heat at a rate of about 1000 watts (1 kW) when operating at 240 volts and consume a current of 4.2 amps. Resistors must be chosen so that they will not overheat and melt in a particular application.

A 1000 megawatt (1000 MW) power station will ideally be able to supply electricity to a million single bar (1 kW) electric heaters while using enough fuel to give it energy at a rate of 1000 million joules per second. The total current supplied at mains voltage (240 volts) will be four million amperes.

Horsepower

The horsepower (hp) as a unit of power is still quite frequently used, being equivalent to 550 ft lbf/s or 750 W (approximately). It is worth noting, however, that the average horse working for lengthy periods can develop only 0.6 to 0.7 hp. James Watt used strong dray horses for short periods in his research a century ago. A 100 hp electric winding engine operating at 1 kV will require 75 amps when pulling 7½ tons up a shaft at a speed of 3 ft/s.

Below: the measurement of horsepower using Regnier's dynamometer, which is a tension-measuring device. Dynamometers measure either power or force: nowadays they usually measure the torque necessary to stop a turning shaft. Horsepower was originally a calculation used by James Watt in designing his steam engines.

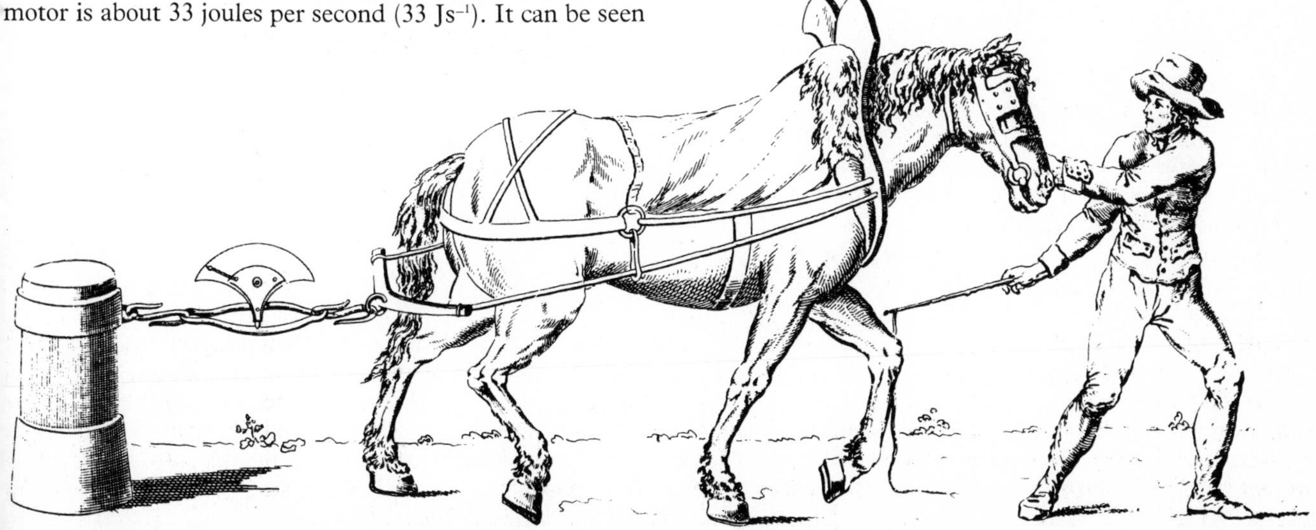

Sir Isaac Newton

On the early morning of Christmas Day 1642, in a Lincolnshire farmhouse, was born the greatest scientific genius the world has known. Isaac Newton was a premature and tiny baby, and the midwives who went off to get medicines for him doubted whether he would be alive on their return. From these undistinguished origins arose the man whose mind laid the foundations of many branches of modern science.

Isaac's father, an illiterate yeoman farmer, had died three months previously. Although the cottage at Woolsthorpe is described as a manor house, and Isaac Newton became in effect lord of the manor, he was not heir to any fortune and he was expected to become a working farmer like his father.

His background did allow him to receive a good schooling in the local day schools and, later, in nearby Grantham. He was by no means an outstanding scholar—he was not very interested in the classical studies of the day, and spent much of his time making working models and studying the world around him. When prompted by his schoolmaster, by such tricks as placing him low in the class, he would make a successful effort to shine at his schoolwork. All his pocket money, however, went on buying tools so that he could make better gadgets.

Newton's mother had married a local vicar and gone to live with him when Newton was only four, leaving him in the care of his grandmother: this may well have had a traumatic effect on him, affecting his later relationships with people.

When he was in his early teens, however, his stepfather died and Isaac had to return to Woolsthorpe to help with the farm. He was of less use than his mother had hoped. When tending sheep, for example, he would get so engrossed with a book, some invention, or in watching the stream, that the sheep would very likely get into the corn. Once, when returning from Grantham, he dismounted for a hill, forgot to remount and plodded home leading the horse, deeply lost in thought. Throughout his life, Newton had the reputation of being a true absent-minded professor.

His schoolmaster recognized the intellect that was within Newton, and persuaded his mother to let him return to school. When Isaac eventually went to university at Cambridge, the farm workers reckoned that was all he was good for, and would never make anything of himself.

Newton's degree was undistinguished, but at Cambridge he met Isaac Barrow, then professor of mathematics, who must have sparked off something within Newton for shortly afterwards the dreamy-eyed boy began to concentrate on his most original work. In his graduation year, 1665, when Newton was 23, bubonic plague broke out in England and he returned to Woolsthorpe to avoid it. It was here that all the influences on him had their effect: he looked back on those two years of seclusion at his mother's cottage as the most significant of his life.

Newton himself was the source of the famous story that while sitting in the orchard on a warm autumn afternoon, the fall of an apple to the ground set him wondering about the nature of gravity. Talking later, he said that he wondered why the apple always fell towards the centre of the Earth, and reasoned that all matter attracts the rest of matter to it.

At Woolsthorpe, too, he performed his experiments on the nature of light, passing white light through a prism to produce a spectrum (Newton's word) of colours. He realized that white light is no more than all the colours seen together, and made many more investigations into the nature of light and optics.

All this time, he was developing his ideas on mathematics, resulting in the discovery of calculus or, as he called it, the 'Method of Fluxions'.

On Newton's return to Cambridge, Barrow stepped down in favour of him as professor of mathematics. One might imagine that Newton would publish the results of his work as soon as possible; actually he seems to have had little desire to publish, and when he did produce his results it was only at the insistence of others.

Newton's introduction into the fully academic life came with his election to the Royal Society in 1672 when he made the first working reflecting astronomical telescope. He gave a paper reporting his works on optics, and at once fell foul of Robert Hooke,* who upheld the wave theory of light as against Newton's corpuscular theory. Actually, each was right in his own way, according to modern quantum theory.

Hooke and Newton quarrelled many times over the years; there were also bitter debates as to whether Newton or the German mathematician Leibniz had invented calculus. The truth is that many of Newton's discoveries were in the air at the time, and would before long have been put forward by others; but Newton's skill and genius tied the loose ends together to produce the final results.

The *Principia*, which describes Newton's Laws of Motion and Gravitation, was published in 1684, and *Opticks* in 1704. By this time Newton, seeking a more prominent situation than that of a mere academic, took on an important post at the Royal Mint, where he played a large part in revising the coinage system.

Newton never married, though there were romantic tales of a childhood sweetheart. He seems to have reserved his attentions for work only. He spent a vast amount of time studying theology, alchemy, and the chronology of ancient civilizations, but his work on these subjects is now practically forgotten. In his own day, as now, his work on physics and mathematics made him an internationally renowned figure. When he died in 1727, he was buried in Westminster Abbey, an honour previously reserved for monarchs, politicians, soldiers, and other such worthies.

Newton's laws

In the course of his pioneering experiments in many branches of science, Sir Isaac Newton discovered several of the fundamental laws of physics. A scientific 'law' is a general statement which can explain the results of a number of different experiments; this generalization can then be used to predict the outcome of other, similar, experiments.

The basic principles of dynamics (the study of how forces act on objects) are summed up in Newton's three Laws of Motion.

The First Law says that *any moving body will continue to move in a straight line and at a constant speed unless it is acted upon by an outside force.* This is not immediately obvious, since on Earth we are used to moving objects eventually stopping, but this is because outside forces, such as friction and air resistance, act on the body to slow it down.

What happens when, as in most practical cases, an outside force does act on a moving body is covered by the Second Law: *if a force is applied to a body, its momentum will change in such a way that the rate of change of momentum is equal to the magnitude of the force.* The momentum is the mass of the body multiplied by its velocity (speed), and so another way of stating the second Law is that *the force on a body is equal to the mass of the body multiplied by the acceleration produced by the force:* Force = mass × acceleration. A consequence of this law is that if the same force is applied to two objects with different masses, the less massive body will accelerate more than the more massive one.

The Third Law states that *for every action there is an equal and opposite reaction.* The action and reaction refer to the forces on two different bodies; for example, the weight of a chair standing on a floor must be balanced by an upward 'reaction' force of the floor on the chair, or else (according to the second Law) the chair would be accelerated towards the centre of the Earth.

Anyone who tries to jump off a stationary toboggan notices an effect of the Third Law: the toboggan begins to move in the opposite direction even before the person's feet touch the ground.

The action of a rocket demonstrates all the laws of motion. A rocket at rest on the ground, or coasting through space with the engine switched off, is obeying the First Law. When the engine is on, the force with which the propellant is ejected from the rocket must be balanced by a reaction force of the propellant on the rocket, and it is this reaction force which drives the rocket forward. These two forces must be equal, and act in opposite directions (Third Law). Since the mass of the rocket is much greater than that of the ejected propellant, the rocket is accelerated to a very much slower velocity than that of the propellant (Second Law).

Newton's Law of Gravitation describes how the gravitational force between any two objects varies with their masses and the distance between them. Each body experiences a force equal to the product of the masses of the two bodies multiplied by the universal Constant of Gravitation, G, and divided by the square of their separation. Newton was unable to explain the origin of gravitation, and Einstein's General Theory of Relativity (1915) proposed that the geometry of space near massive bodies is altered, so that the quickest distance (the *geodesic*) between two points is not a straight line. By substituting the word 'geodesic' for 'straight line' in Newton's First Law, Einstein was able to incorporate gravity in the First Law of Motion. In spite of relativity, however, the Second and Third Laws are still, three hundred years after their formulation, fundamental to modern science.

Newton's investigations of the cooling of a hot body led to his law of cooling: *the rate of cooling is proportional to the difference in temperature between the object and its surroundings.* The rate of cooling is measured by the rate at which the temperature of a body falls; so a body at a high temperature relative to its surroundings will initially cool fast, and its temperature falls rapidly. As its temperature decreases, however, the cooling becomes slower, and so the temperature falls less rapidly. This type of behaviour is known as an *exponential decay* and strictly speaking, although the temperature of the body becomes closer and closer to that of the surroundings, it will never become exactly the same.

Below: Newton reasoned that the same force which caused the apple to fall to earth kept the planets in their orbits. This leap in thought led to the Apollo flights to the moon, 300 years later.

POTENTIAL

Potential is a concept used in the discussion of any 'conservative' field of force. By conservative field it is meant that energy is conserved. In moving an object through the field, the only energy gained or lost is through the change in potential energy—for example, there is no friction in the movement whereby heat could be generated. Another way of defining a conservative field is that the energy required to move an object in the field is independent of the path taken. Potential theory can thus apply to a mass in a gravitational field, a charge in an electric field, a pole in a magnetic field, and a nuclear particle in a nuclear field. These situations are treated identically as far as potential energy is concerned, so it is only necessary to consider one of them, for example, gravitational potential.

As the name indicates, the potential of an object is the energy it can potentially release as it moves to a 'lower' energy level. Thus work is done on the 'monkey' of a piledriver to raise it higher in the Earth's gravitational field; when the monkey is released it moves to a lower energy level, and dissipates the difference by doing work on the pile—applying a force and causing it to move to a certain distance. (Much of the energy is of course 'wasted' as heat and sound.)

Clearly to obtain more work from the monkey at a given site, the engineers could increase the height from which it is released or alternatively increase its weight. The gravitational potential energy in such circumstances may be found by multiplying the weight of the object by its height. Thus a weight of 100 newtons (25 poundsforce) at a height above ground of one metre (three feet) would have a potential of 100 newtons-metre (75 foot-poundsforce). That much energy will be released and that much work done, when the object falls.

Implicit in this discussion has been the fact that potential is taken to be a relative term (as in electricity one speaks of 'potential difference'). True, or absolute, potential must be referred to a 'zero' level, and this is generally at an infinite distance from the source of the field. In practice physicists and engineers will choose an arbitrary zero reference level to suit their particular purpose. Where gravitational

Below: the Victoria Falls, Rhodesia. At the top of the fall, the water has maximum potential energy in the Earth's gravitational field. After it goes over the edge it has kinetic energy, which might be used, for example, to turn turbine blades and generate electricity, or to turn a mill to grind corn.
Right: the rock has gravitational potential.

potential is concerned this is often the Earth's surface, which is the case too with electric potential (the 'earth', or ground', is said to be at zero potential).

The concept of the potential at a point in a field is also commonly encountered. In a gravitational field it is the work required to bring unit mass—one pound or one kilogramme—to that point from an infinite distance. In the case of electricity it is the work or energy involved in bringing unit charge—one coulomb—from infinity to the point in question; the special unit used here is the volt.

As well as the examples already discussed, potential is found in various other kinds of system. A clock spring, when wound, has potential energy, being able to drive the mechanism as it uncoils—this is mechanical potential. A mixture of hydrogen and oxygen can do a great deal of work when ignited—this is chemical potential.

INERTIA

It is a matter of everyday experience that it requires effort to start objects moving, to change their direction of movement, and to stop them again. This resistance to changes in speed and direction of movement is called inertia. Even if all resisting forces such as friction were absent, the inertia of, say, a truck would limit the rate at which an applied force would accelerate it, or speed it up. The truck's inertia becomes apparent in the force that its brakes must apply to slow it and that its tyres must exert on the road. If free of all such forces the truck would keep moving in a straight line at constant speed indefinitely. This generalization about the behaviour of moving bodies is the first of Newton's Laws.

The inertia of a body which is not rotating depends simply on its mass. Although the *weight* of an object varies depending on the strength of the gravitational field it is in, its mass, and consequently inertia, will remain unchanged wherever the object is; its inertia will still manifest itself as a resistance to any applied force.

It is inertia that keeps fans, spin driers, sirens and aircraft engines turning after their power supply has been disconnected. Inertia in respect of rotational motion is given the name *moment of inertia*. This depends not just on the mass of a body, but on the distribution of that mass as well. A body's moment of inertia can be increased by placing its mass farther from its axis of rotation. For example, if two wheels have the same mass but different diameters, the larger wheel will roll downhill more slowly because of its larger moment of inertia. If a steel disc is melted down and recast as a wheel with the bulk of its mass at its rim, its moment of inertia will be increased. But there will be no change in the body's resistance to non-rotational motion. How hard it is to throw a discus depends on its mass, not on its moment of inertia.

Above: this turntable has maximum inertia for minimum mass; the aluminium platter has brass weights around its circumference to damp out speed variation.

Every moving object is a store of *kinetic* energy, or

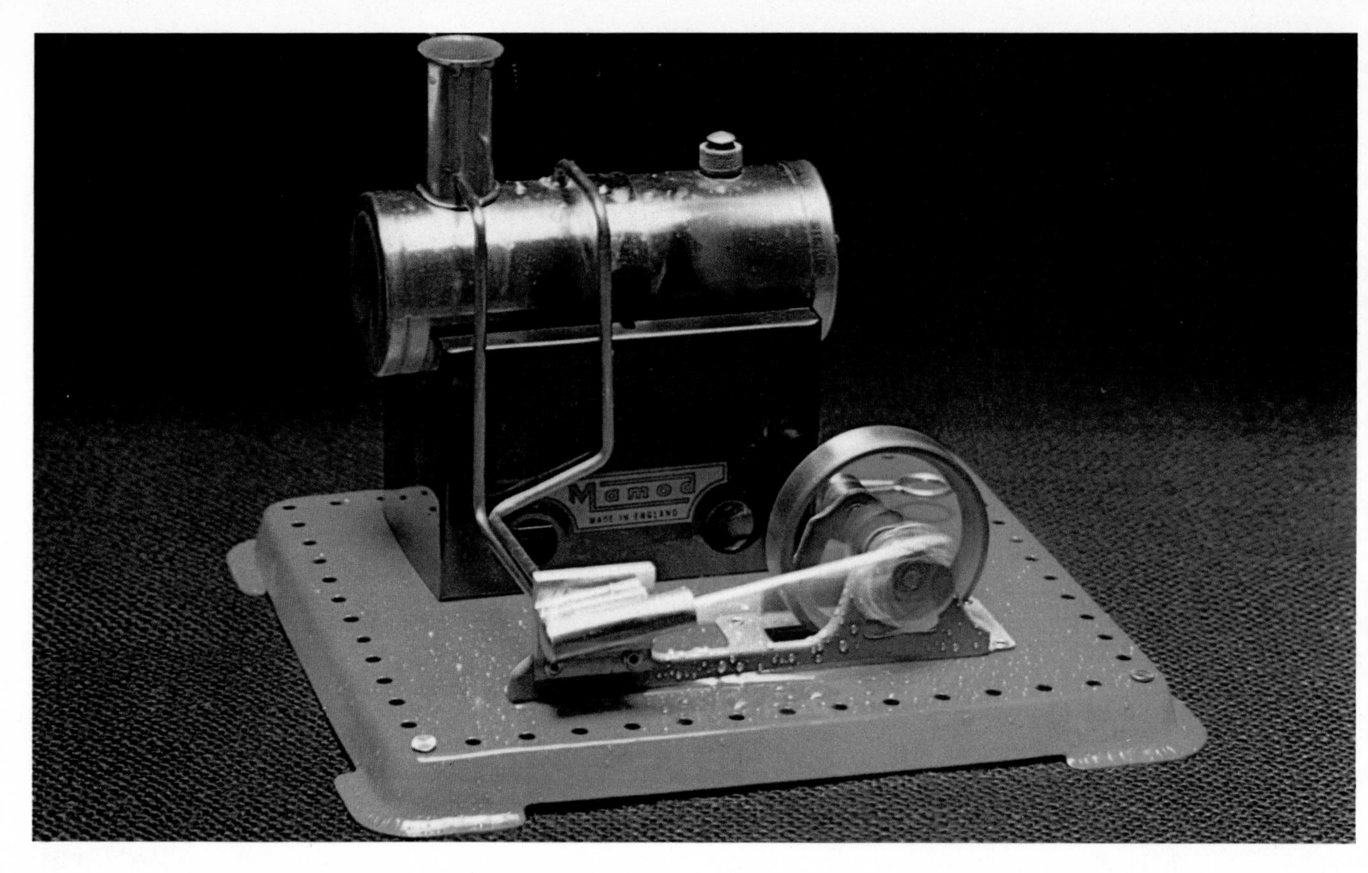

Above: without a flyweel, this model steam engine would stick at the end of its first stroke. It needs a spin to start it off, but then the inertia in the flywheel keeps it going. For centuries men dreamed of perpetual motion machines suggested by devices such as flywheels.

energy of movement. A more massive object moving at a certain speed has more energy than a less massive one at the same speed.

In the case of a rotating body, kinetic energy depends on moment of inertia and rate of rotation in revolutions per second. A classical illustration of the effect of changes in moment of inertia is given by a spinning ice skater. By throwing out his arms he can increase his moment of inertia (by redistributing his mass farther from his axis of rotation). Since his kinetic energy is unchanged, his rate of rotation falls to compensate for the increase. When he pulls his arms in again, his moment of inertia is reduced, and he spins faster.

Uses of inertia

Many kinds of engines deliver power unevenly through their working cycle. The moment of inertia of a flywheel can be a means of smoothing this energy supply. For example, the crankshaft of a car receives impulses from the firing of the engine cylinders. When the crankshaft is disengaged from the road wheels—that is, when the car is in neutral gear—it would be liable to turn jerkily. A flywheel bolted to the crankshaft keeps it turning smoothly.

Conversely, a smoothly working engine may power a process that requires energy discontinuously. Looms, mechanical hammers, shears and presses all require bursts of energy that electrical motors in particular are ill-fitted to deliver. In these cases the motors drive flywheels, and the flywheels are coupled to the machines by engaging a clutch when needed. The greater the moment of inertia of the flywheel the smaller the drop in speed of revolution caused when it delivers a burst of energy.

Frames of reference

The inertia of a massive pendulum was used by the physicist Foucault to demonstrate the rotation of the Earth. While the Earth turned beneath it, the pendulum tended to keep swinging in the same direction in space. The result was that the pendulum's line of swing over the ground seemed to change in the course of a day. The same effect gives rise to Coriolis force, which makes cyclones blow in circular motions.

Even if the stars were never visible from the Earth's surface we could infer the Earth's rotation by observing the behaviour of pendulums and of the winds. It might seem easy to go farther. Even if the stars did not exist and the Earth were alone in space, surely the existence of Coriolis forces and other inertia effects would show up the Earth's rotation?

Such an idea was a scandal to the physicist Ernst Mach, who worked at the turn of the century. If no distant celestial objects existed by which a standard of rest could be defined, it seemed to Mach that the idea of motion of the Earth would be meaningless. This led him to suppose that if the stars did not exist, the effects of the Earth's rotation that we observe would not occur. In short, the inertia of objects on the Earth, however minute, is caused by the existence of the rest of the Universe. Astronomers have found that even though our own Galaxy or star system is turning, the inertial frame of reference against which we measure such rotation depends on even more distant matter, at the limit of the observable Universe. If the celestial bodies were to vanish, we should find ourselves living in a world of objects as light and as unresisting to pushes and pulls as thistledown.

Bizarre as these ideas sound, they bore fruit in the work of Einstein. In ways not dreamed of by Mach, Einstein's general theory of relativity established a link between the inertia of each piece of matter and the distribution of all other matter in the Universe.

DYNAMICS

Dynamics is a branch of physics concerned with moving bodies—their direction, speed, momentum and energy, and the inter-relation of these quantities. It therefore represents one side of the subject of mechanics, the other being statics—the study of stationary bodies, and the forces on these, in a stable, non-moving, situation. The design of a bridge, for example, requires the use of statics to determine its structural stability. Dynamics, on the other hand, being concerned with movement, seeks to determine quantitatively the effects of a force on a body's motion and as such is closely related to the subject of applied mathematics. Indeed, once the relevant laws of physics have been employed to analyze the given situation and establish the mathematical expressions of these laws, the problem becomes largely one of the manipulation and solution of equations.

Kinematics and kinetics

Dynamics is split into two subsections: kinematics and kinetics. *Kinematics* deals with the mathematical description of the body's motion such as its speed and velocity, and does not actually touch the physics of the situation. Once this is done, an analysis from the point of view of the laws of physics that govern motion is performed: this is the *kinetics* part of the investigation.

Sir Isaac Newton's laws of motion formulated in the seventeenth century and developed by later physicists and mathematicians were spectacularly successful in their explanations and analysis of many problems, from the motion of planets around the sun to the behaviour of tiny particles of dust. The discovery of Neptune in the mid-nineteenth century, for example, was not due to improvements in the optical properties of telescopes, but to the application of dynamics.

A dynamical analysis of the solar system had shown that the observed motions of the known planets were at variance with what was predicted theoretically. It was then realized that the existence of an eighth planet was necessary as the deviations could only be explained if there was a distant planet whose gravitational attraction was distorting the orbits of the others. The analysis even predicted where to look for this planet, its size and some details of its orbit around the sun. It was located within a couple of degrees of the calculated position by Professor Galle in Berlin in 1846, following independent calculations by Leverrier in France and Adams in England.

Scalar and vector quantities

In dynamics there is found first a precise definition of all the properties that bodies exhibit because they are moving—these properties are called *dynamical variables* and include speed, velocity, acceleration, momentum and energy.

Speed is defined as distance travelled per unit time and is

Right: the lift-off of the Apollo 11 moon mission in July 1969. A tremendous amount of thrust is needed to overcome the inertia of the rocket and get it moving fast enough to escape the Earth's gravitational field.

a *scalar quantity*, meaning that it has a magnitude expressed only in units of speed such as miles per hour or metres per second. Many dynamical variables, however, are *vector quantities* which means that they must be assigned a direction in space as well as a magnitude.

Velocity, for example, is defined as speed in a particular direction and is a vector quantity. Thus two cars may have the same speed if they cover the same distance in the same time, but will only have the same velocity if they are travelling in the same direction (that is, parallel). Momentum is also a vector quantity because it is defined as mass multiplied by velocity and the direction of this vector is the same as the velocity vector involved in its calculation.

The momentum of a body is a measure of how difficult it is to bring it to rest. Consider two bodies, one with twice the mass of the other but the smaller mass travelling with twice the speed of the larger. They are equally difficult to stop because the doubled mass of the slower body compensates for the higher speed of the smaller mass—each body has the same momentum.

Vectors are of fundamental importance because a body will only change its line of travel, or more generally, its momentum, if it is forced to do so. Just how large that force must be and exactly how quickly (and by how much) it will alter its direction is basically what dynamics is all about.

The concept of force is central to the whole theory of dynamics. Newton's laws of motion state that a body's momentum (and therefore its speed and direction of travel) can only be changed by the application of a force. The rate of change of momentum with time is determined by the magnitude of the force and the direction of change by the direction in which the force acts. Force is therefore also a vector quantity. Energy, work and power are all scalar quantities because they cannot be related to a specific direction. Energy, for example, can be harnessed to do work in any direction.

Rotary motion

With linear motion the force acting on a body is equal to the mass of the body times its acceleration (Newton's laws). With a rotating body there is a similar relationship between the twisting force, or *torque* (from the Latin verb 'torquere' —to twist), and angular acceleration; torque is equal to the moment of inertia of the body times the angular acceleration.

Dynamics and modern physics

Like so many branches of physics, dynamics had to be substantially revised in the light of modern discoveries. To explain properly the newly observed phenomena that occur on the microscopic level, such as the collisions and interactions between electrons and nuclei, Newtonian (or as it is now known, classical) mechanics is inadequate and quantum mechanics is required. Also, the strange behaviour of bodies moving with speeds comparable to that of light cannot be explained by classical mechanics and relativistic mechanics becomes important. Practical applications on the scale and complexity of those found in most real-life engineering projects, however, need not resort to such advanced approaches.

HEAT AND THERMODYNAMICS

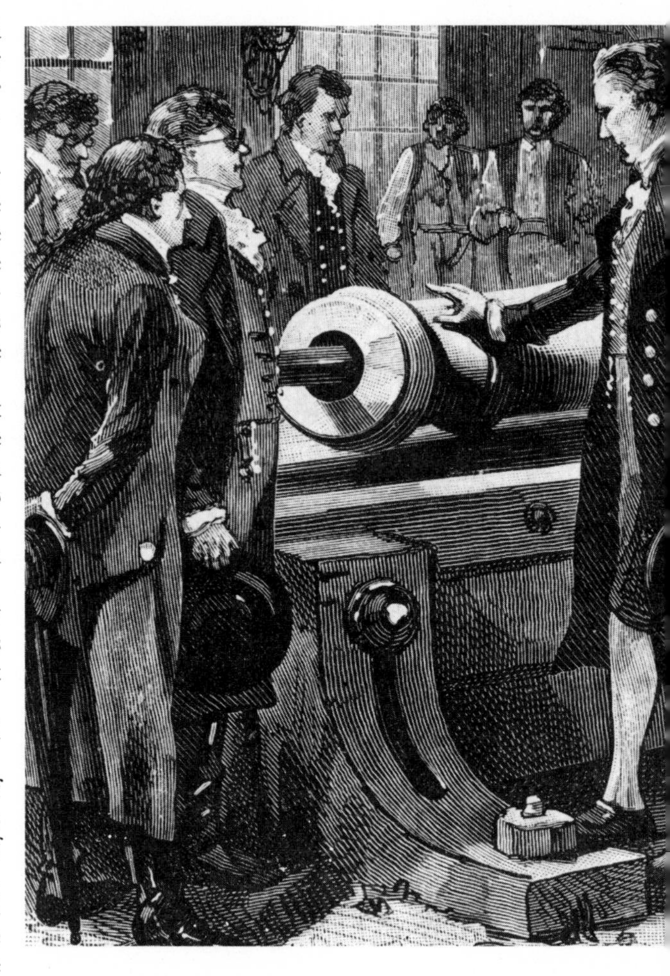

From the practical problems of building an efficient stear engine came a science which deals with such things as th course of chemical reactions, the absolute temperatur scale used in physics—and which even predicts the ulti mate fate of the Universe.

The science is called *thermodynamics*, which literall means 'power from heat', and it began in the early 19t century. The basis of thermodynamics is the study of th internal energy stored in a physical system, which could b for example, a gas, an electric battery cell or a stretche wire. The internal energy is dependent on certain coordi nates of the system, which are pressure and volume (for gas), voltage and charge (for a cell) or tension and lengt (for a wire); in all cases the temperature is also an importan coordinate. When the coordinates of a system are altere there is usually an exchange of mechanical energy (work and heat with the surroundings, and it is these exchange which thermodynamics mostly deals with.

In making calculations to do with heat another importan quantity is the *specific heat* of a substance. It is found tha weight for weight, a substance such as copper is easier t warm up than water. This is due to their differing values o specific heat—usually defined as the amount of hea measured in calories, required to raise the temperature o 1 gramme of the substance by 1°C. Knowing the specifi

Left: Count Rumford was an American of British parentage who was knighted in 1784. He was also made count of the Holy Roman Empire for his work in Bavaria as Minister of War. His study of the heat produced in the boring of cannon led to his theory of the mechanical equivalent of heat. He saw that the amount of energy converted could be crudely estimated by surrounding the cannon with water and measuring the rise in temperature, and also theorized that heat might be due to molecular motion. Below right: a heat exchanging device for cooling intergrated circuits. Water is sealed in pipes at pressure so that it is always near boiling temperature; a small rise in heat is enough to produce vapour, which flows to the cooler region and condenses back into water.

heat of a material, the amount of heat needed to raise the temperature of any number of grammes of it by any number of degrees can be easily calculated by scaling up the specific heat.

In engineering the objective is to produce useful energy as efficiently as possible, and the theoretical efficiency of any type of engine can be calculated from the First and Second Laws of thermodynamics. In practice the efficiency is inevitably less because of friction and heat losses by conduction through the engine.

First law

The nature of heat was not firmly established until the work of J P Joule in the 1840s. Until then it had been generally believed that heat was a light fluid (*caloric*) which permeated all substances, and that when two bodies were brought into contact this fluid flowed from the hotter to the colder body until the temperatures were equal. The caloric idea had been questioned by Count Rumford, who observed that when cannon were bored the outflow of heat continued for as long as the boring lasted, and that if all the heat were put back it would be sufficient to melt the cannon. He concluded that the work of boring must somehow be converted into heat. Joule measured the heat produced by a certain amount of mechanical work, and showed that it was always the same, regardless of how the conversion was done. The modern value for this quantity, known as the mechanical equivalent of heat, is 4.18 joules of mechanical work being equivalent to one calorie—the amount of heat required to raise the temperature of one gramme of water by one Celsius degree (°C). In a modern experiment it is common for quantities of heat to be measured directly in joules, as the traditional units of heat are actually superfluous.

Joule's work showed that heat is simply a form of energy. The important conclusion which follows from Joule's careful experiments is that in any closed system the total amount of energy is always conserved. This statement is referred to as the Law of Conservation of Energy, or as the First Law of thermodynamics. One consequence of the law is that perpetual motion machines are impossible to construct, as they would require continuous creation of energy to keep the machine turning at a constant speed while energy is being continuously drained away as heat by friction at the bearings.

The Carnot engine

Even before the law of conservation of energy was fully understood, various types of engine for converting heat into more useful forms of energy had been invented. These were first used in the early 18th century for pumping water from mine shafts, and they were later adapted for many other purposes, including the railway steam engine. All these engines were very inefficient, converting less than 10% of the heat energy into useful work, and this prompted the French physicist Sadi Carnot in 1824 to investigate theoretically the efficiency of heat engines. He considered

the action of a 'perfect' engine which is perfectly insulated and frictionless, absorbs heat at one temperature and rejects it at a lower temperature, and converts the maximum possible amount of heat into useful energy. It includes a 'working substance' which is continuously recycled (in practice this could be steam), and the engine is completely reversible. If operated in reverse it could use the mechanical energy supplied to make heat flow 'uphill' from the lower temperature reservoir to the higher; this is the principle of the heat pump.

The Carnot engine is more efficient than any engine which could actually be built, but Carnot found that even his theoretical engine was not 100% efficient. Its efficiency (the fraction of heat converted to work) is independent of the working substance, depending only on the *temperature difference* between the two heat reservoirs, and it is always less than 100% unless the lower reservoir is at the absolute zero of temperature (—273.16°C). Lord Kelvin realized in 1884 that the Carnot engine could be used to define an absolute temperature scale which was independent of the thermal properties of any particular substance, and this, the *Kelvin* scale, is still the fundamental temperature scale in physics. Temperatures below absolute zero must be impossible to obtain, because otherwise the efficiency of a Carnot engine would become greater than 100% implying that energy is being created and contradicting the First Law. More recently it has been shown that in real experiments the temperature cannot even be reduced to exactly absolute zero—although it can be approached very closely.

Second law

The most important result of Carnot's analysis of heat engines is that heat can only be changed into useful energy if it 'flows' between reservoirs at different temperatures. It is impossible to take heat from just one reservoir at a particular temperature and convert it into work. This statement is one expression of the Second Law of thermodynamics (another commonly found statement is that 'heat will only flow spontaneously from a hotter to a colder body') and it means, for example, that it is not possible to power an ocean liner just by using the heat energy in the ocean.

It is the difference in temperature between the Sun and the Earth which allows the production of the Earth's energy resources (apart from tidal and nuclear power), both in the kinetic energy of winds, and more important, in the chemical potential energy stored in plants as a result of photosynthesis. The Earth would be heated up to the Sun's temperature of 6000°C if it were not radiating heat into the even colder reservoir of space, at a temperature of —270° C. Some physicists have argued that because all useful energy is ultimately converted into heat by non-reversible processes like friction, and because the conversion of heat to work is not only inefficient but tends to bring all heat reservoirs to the same temperature, the Universe will eventually reach a final state, the 'heat death', in which all astronomical bodies have the same temperature

Above top: a spectacular example of increasing entropy. The sequence of pictures was taken in 1958 and covers about twelve seconds; it shows the destruction of Ripple Rock, a shipping hazard in Seymour Narrows, British Columbia. In this instance the increased disorder in the universe is very unlikely to reverse itself.
Above: entropy on a smaller scale. Sugar in tea spreads throughout the solution by diffusion, hastened by stirring. Here, potassium permanganate crystals diffuse through water.

and no more useful energy can be produced.

Statistical mechanics

A rather different approach to thermodynamics is to consider the motion of the individual molecules in the system (the *microscopic* approach). Heat energy is simply the total random kinetic energy of all the individual molecules, so that heating a gas increases the velocities of all the molecules; while in a solid, where the mean position of each molecule is fixed, the process of heating increases the vibration energy of each molecule about its mean position. In the example of a dropped brick, the total kinetic energy of the molecules is the same just before and just after the collision with the ground, but in the former case all the molecules are moving in the same direction, so that the brick as a whole moves; while after the collision the motions occur in random directions and the brick is stationary, but warmer.

The explanation of *macroscopic* (large scale) properties such as pressure and viscosity in terms of collisions of molecules regarded as miniature billiard balls is known as *kinetic theory*, and it explains the observed gas laws. To explain thermodynamic results, however, a more elaborate theory which describes the molecules in terms of their total energy is required. This is *statistical mechanics*, and it provides a way of predicting the behaviour of many physical systems from the properties of the particles comprising it.

In statistical mechanics the probabilities of different energy states of the system are calculated, because in any reaction the system will tend to change into the most probable state. The likelihood that the individual molecules of the brick will be vibrating in different directions is very much higher than that they should all be moving in the same direction, so it can be predicted that the falling brick (an ordered motion of molecules) will tend to become a stationary warmed brick (disordered molecular motion). On the other hand, the chance that when a brick is heated all the molecules will move in the same direction at the same time is so small that it can be safely predicted that a heated brick will not jump off the ground spontaneously. This result is exactly the same as would be derived from the Second Law, in that ordered energy (work) can be converted into heat (disordered energy), but not vice versa. Statistical mechanics can thus be regarded as a microscopic explanation of thermodynamic results.

Sadi Carnot

Sadi Carnot was a French physicist who founded the science of thermodynamics, which literally means 'heat movement'. His early death from cholera at the age of 36 was a great loss to science in the 19th century.

Carnot came from one of the most prominent families of his day. He was the son of Lazare Carnot, the French minister for war, and elder brother of Hippolyte Carnot, a French politician whose son, Marie Francois Sadi Carnot, was President of the French Third Republic. Hippolyte's second son, Mari-Adolph, was a mineralogist after whom the uranium ore carnotite is named.

Carnot was educated at the Ecole Polytechnique, noted for its science teaching. Although he was nominally trained to be a military engineer, his main interest was in the harnessing of heat energy, as in steam engines. Carnot has been called the most profoundly original French physicist of his day—a reputation based on his sole publication, a book called *Reflexions sur la Puissance Motrice de Feu*. In this book, published in 1824, Carnot set out to find whether the motive power of heat is unbounded, and whether possible improvements in steam engines have a limit or whether these improvements may be carried on indefinitely.

Although steam engines were becoming common, no one fully understood the principles on which they worked. The engines were also very inefficient—none of them turned more than 10% of their heat energy into work. Carnot found that there was a relationship between temperature and work done in the engine, which was a glimpse of what was later to become the first law of thermodynamics: that energy is never destroyed, only changed from one form to another.

More importantly, he found that the efficiency of an engine depended on the temperature difference between the heat source (in the case of a steam engine, the boiler) and the heat sink, or receiver (the condenser for the steam engine). Carnot saw that the boiler and condenser were two essentials for a heat engine; without the temperature gradient that they gave, the engine would not work at all. The third and last essential was a working fluid to transfer the heat, which in this case was steam.

Carnot reached his conclusions by visualizing an ideal engine in which the processes of heating steam—using it to push a piston, and letting the piston return by condensing the steam—all took place in a repeatable cycle. This so-called Carnot cycle, which is theoretically reversible, can never be obtained in practice because in a real engine some energy is always lost through friction.

Carnot showed that the nature of the working fluid does not matter, and neither does the way the temperature drops between the heat source and the heat sink. The maximum efficiency of the engine is governed purely by their temperature difference. This work is the basis for the so-called second law of thermodynamics, which in its simplest form states that heat only flows spontaneously from a hotter body to a colder one. Although other scientists, such as Clausius and Kelvin, later developed Carnot's work, his was the first clear analysis of the processes occurring in heat engines.

Left top: Sadi Carnot, like many early experimental scientists, was educated in military engineering.
Left: the heat regenerator of a modern Stirling engine. The engine works by circulating a heated fluid from a hot to a cold section and back again, pushing a piston.

NTROPY

The entropy of a system is a measure of its disorder, and in ny change affecting a closed system the entropy can only ncrease, as in the above example. This is another way of tating the Second Law of Thermodynamics (due originally o Clausius, who first introduced the idea of entropy in 865), and it is interesting because it defines the direction of time. Most of the laws of physics would be equally valid f time ran backwards; for instance, a film of a swinging pendulum would be indistinguishable from the same film un backwards, but if a process involving an entropy change occurs (such as the stopping of the pendulum by air resistance or friction) it is immediately obvious which way the film should be run. Entropy can also be described as the changing of useable energy into non-useable energy.

In 'normal' time, events proceed in the most probable direction—of increasing entropy—and this can be used to define the forward direction of time. In the falling brick example, the probability of the brick jumping off the ground is so small that there is no doubt about which way a film of it should be run.

Entropy also determines the course of a chemical reaction. Most reactions give out heat, but this is not universally true: some reactions will absorb heat if the entropy of the system increases in the process. One example s the dissolving of certain chemical salts in water, where the entropy increases as the ordered crystal becomes a relatively disordered solution, and heat is absorbed from the surroundings to enable the reaction to proceed. A fall n temperature of several degrees can easily be produced in this way.

It is sometimes said that life contradicts the law of entropy increase, because in biological evolution life forms have become more complex and better ordered. The 'biosphere' is not a closed system, however, as it does not include the source of energy, the Sun; and when the total system is considered it is found that the increase in the Sun's entropy is much greater than the decrease due to evolution on the Earth. The action of life is therefore only to slow down, not to reverse, the universal increase in entropy, which may culminate in the eventual heat death of the Universe.

FLAME AND FLASHPOINT

Although a precise definition is elusive, flame is generally recognized as visible combustion which is associated with the release of heat at high temperature. A flame involves chemical processes, in the form of exothermic (heat releasing) reactions which promote molecular excitation and produce light. It also involves the physical processes of transfer of both matter and energy. The heat release arises because of the changes in chemical bonding. Each molecule of fuel and oxidant is made up of atoms bound together by forces of an electrostatic nature which represent a certain level of energy within the molecule. When this same quantity of energy is supplied to the molecule, the bonds are broken, and the atoms become free to re-arrange themselves and bind together into a different pattern, releasing their excess binding energy in the process. The total energy of the product gases is less than that of the fuel and oxygen before combustion, hence the difference in energy is released as heat and light within the flame.

The above combustion equation represents the initial and final materials only; the combustion reaction itself is made up of continuous series of steps with the momentary formation of a whole variety of products of partial oxidation which are themselves unstable and lead to further reaction steps. The vibrations of these excited partial products largely determine the colour of the light transmitted from the flame, and in some cases flame colour can be used to identify the material being burned.

Right: the flame of a gas cigarette lighter has a low premixed zone (blue) where the air for combustion has been drawn in through apertures near the base of the flame, and an upper diffusion zone (yellow) where the air for combustion is taken from the surrounding atmosphere. Far right: a type of burner which produces a large hot premixed flame. Air is drawn in through the holes visible in the barrel of the burner, and is mixed with the gas as it passes up to the mouth.

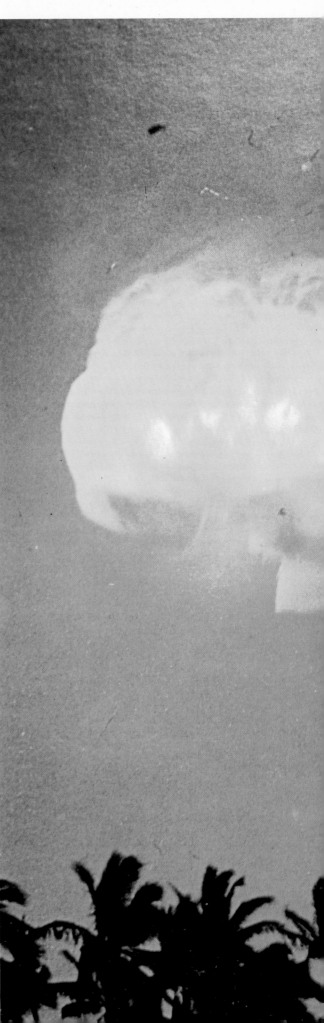

Types of flame

Flames are usually classified as either premixed or diffusion. In the former case, the fuel and oxidant are mixed beforehand at ambient conditions, and are then introduced to the flame where they burn rapidly and, if the mixture ratio is suitable, completely. In the diffusion flame, on the other hand, the fuel meets the oxidant only at the flame itself and the rate of combustion at the interference is then largely controlled by the rate of physical mixing, rather than the rate of chemical reaction.

Both types of flame can be demonstrated with a bunsen burner. If insufficient air for complete combustion is induced by adjusting the air supply to the gas jet at the base, the gas-air mixture burns with a conical premixed flame which, being fuel-rich, gives rise to an outer diffusion flame due to subsequent mixing with the atmospheric air. In fact, the two flames may be drawn apart by means of a Smithell's separator, which consists essentially of a glass tube fitted over the top of a Bunsen burner. A wick in a candle, or a wick fed from a liquid fuel supply, also supports a flame of the diffusion type, the combustion heat released being sufficient to promote a continuous supply of fuel vapour from the wick surface.

The most common types of fuel in general use comprise compounds of hydrogen and carbon, sometimes with oxygen, in gaseous, liquid or solid states. Their premixed flames tend to be blue, owing to the final phase of combustion of carbon monoxide to carbon dioxide, and their diffusion flames tend to be yellow, luminous and sometimes smoky, owing to radiation from carbon particles in the flame. Free carbon is formed by cracking, a disruption of the fuel molecules at high temperature in the absence of sufficient oxygen for immediate combustion. Highly radiant flames are required in heat transfer devices, such as boilers and furnaces, but are undesirable in work transfer devices, such as heat engines, where the energy is required to remain in the gas stream rather than be transferred to the walls.

Flame propagation

A flame propagates through unburned mixture by pro-

Examples of entropy and combustion: far left is a factory fire in Maryland. Old factory buildings are often poor fire risks; such fires are difficult to fight. In the centre is a hydrogen bomb explosion; fusion is the same process that occurs in the Sun. The hot gases rush upward until they reach an equal atmospheric pressure; a non-nuclear explosion of the same size would produce the same effect.

cesses of transfer of both heat and 'active radicals'; these are energetic fragments of molecules which are capable of energizing and triggering off combustion reactions in adjacent layers of fresh mixture. When flames propagate smoothly they are described as *laminar*, and the maximum laminar velocity for most hydrocarbon type fuels is remarkably low, being only about 1½ ft/s (0.45 m/s). This is increased by initial heating of the mixture, and by breaking up and distributing the flame front by means of turbulence.

Ignition

In order to initiate flame, it is necessary to supply sufficient energy to break the chemical bonds in the fuel and oxidant molecules, as outlined earlier. When this energy is provided by means of a pilot flame or spark, the ignition is said to be forced and the mixture of fuel vapour and oxidant is described as flammable. In the case where a liquid fuel is heated in air to produce just enough vapour above the surface to make the vapour-air mixture flammable, the temperature of the liquid fuel is known as the flashpoint of that fuel. This is important when considering the use of many hydrocarbon fuels and solvents, and standard tests have been devised for flashpoint determinations.

For most liquid hydrocarbon fuels of practical interest, the vapour-air mixture just becomes flammable at a fuel vapour concentration of about 1% by volume. Although this fuel concentration is common for all these fuels, the temperature at which it is reached depends upon the volatility of the fuel. Hence, the heavier and more complex fuels are less flammable, since they require heating to higher temperatures before they can be ignited.

When the ignition energy, however, is supplied as heat or pressure only, in the absence of a pilot flame or spark, the fuel and atmospheric oxygen are able to react spontaneously throughout the mixture. The temperature of the vapour-air mixture is known as the spontaneous ignition temperature (T_{sp}) of the fuel, and the lighter and simpler fuels exhibit a higher T_{sp} because their molecules are compact and more able to withstand the thermal agitation that leads to bond breaking. Hence, the heavier more complex fuels are more ignitable, since they ignite spontaneously at lower temperatures and pressures.

Flame temperature

The temperature reached within the flame is the result of a number of chemical and physical factors. The heat released is a known chemical function of the combustion process (the calorific value of the fuel mixture). If no heat is lost to the surroundings, all this combustion heat is absorbed by the product gases, and the resultant temperature thus depends upon their known physical property of specific heat—that is, the heat required to raise the temperature of unit mass of each product by one degree.

In view of the physical as well as the chemical influences upon flame temperature, therefore, the values of combustion temperature do not necessarily follow those of the calorific value of the fuel. This explains why acetylene is used for gas welding in preference to methane for example, since it burns at a higher temperature even though its calorific value is lower.

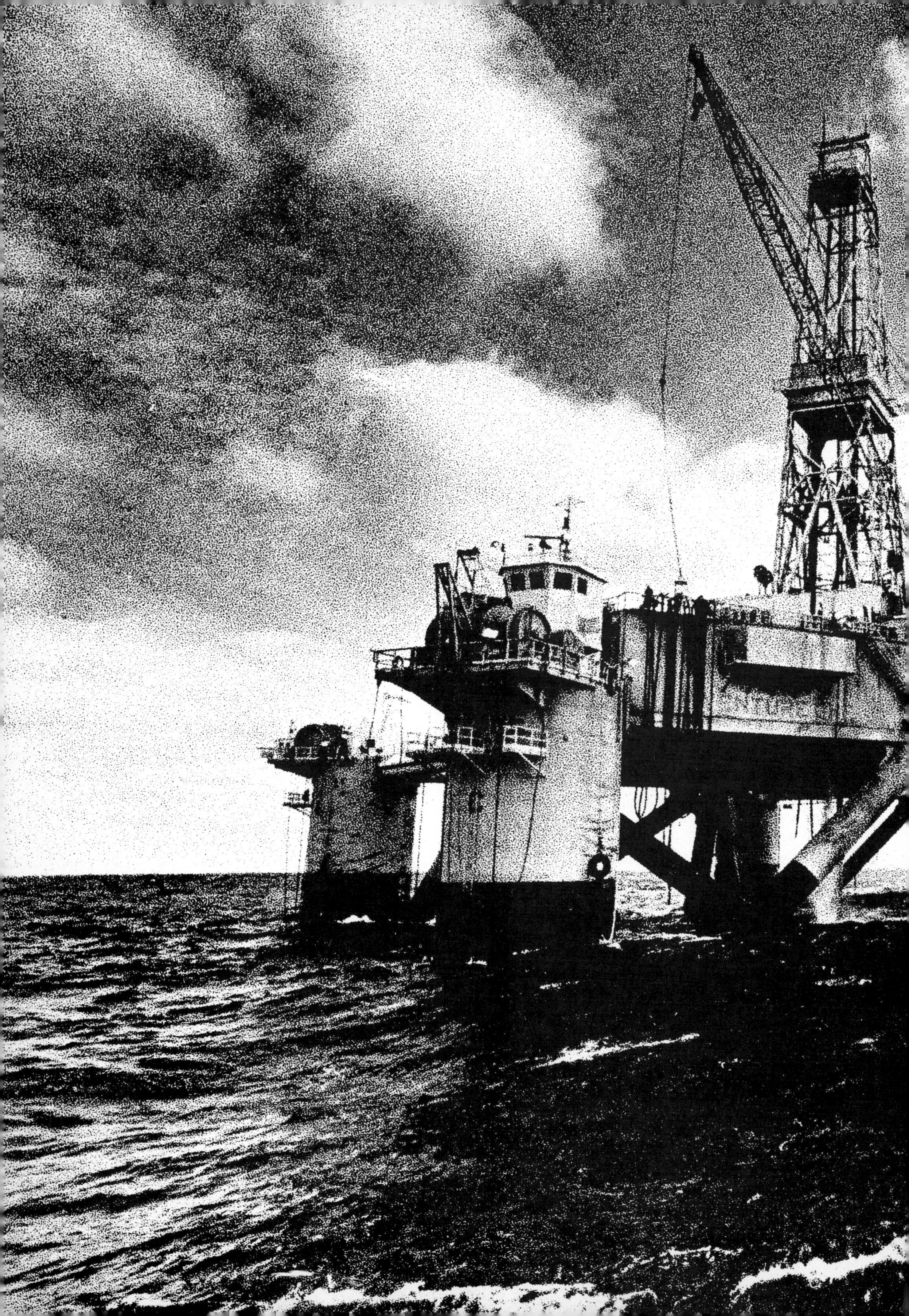

ENERGY SOURCES

ENERGY REQUIREMENTS

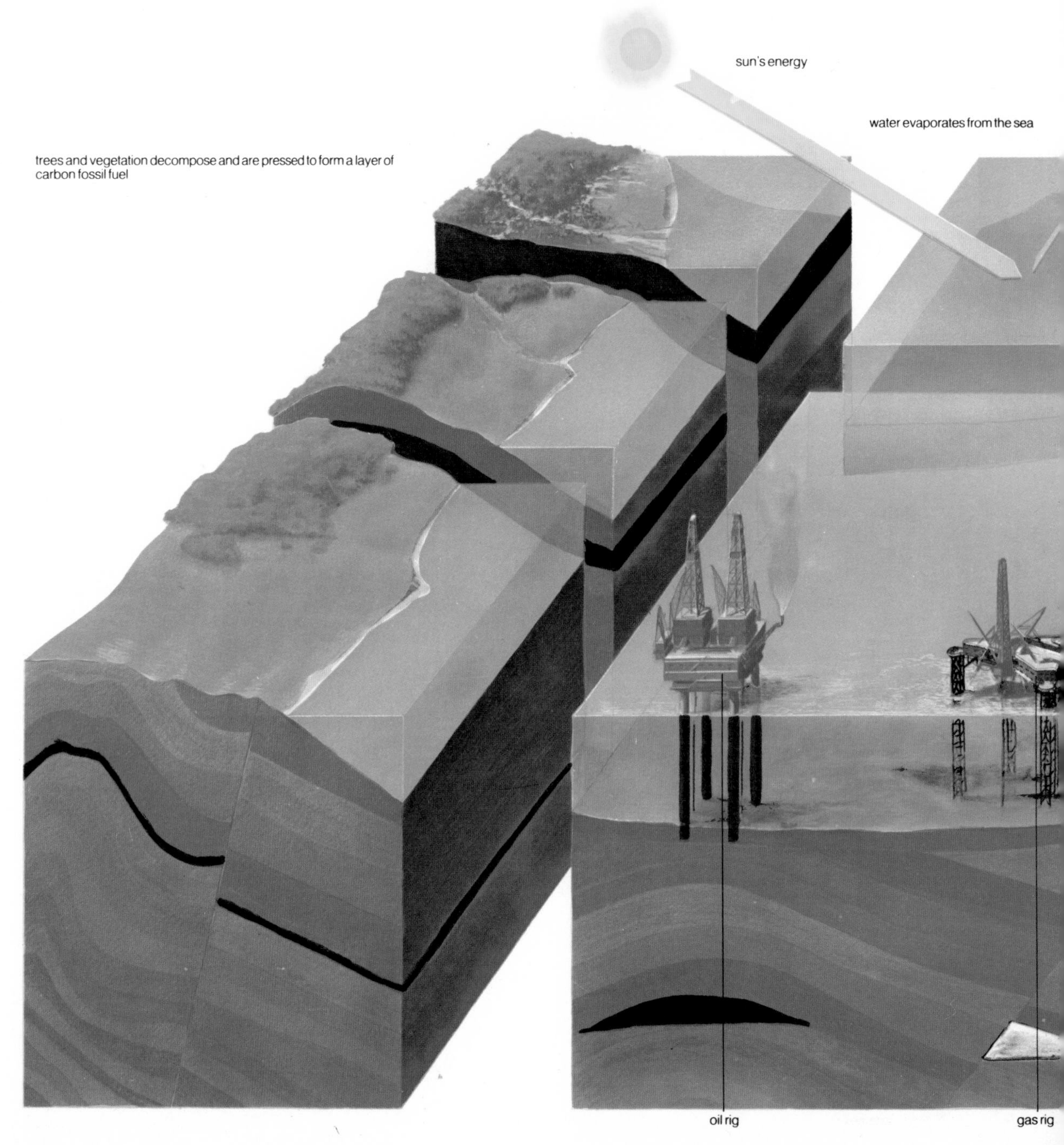

Man's energy consumption has steadily grown from the fires of his primitive ancestors to the modern intensive use of energy in the industrialized world. Only recently has there been any suggestion that there may not be enough energy available to meet the requirements, which now increase at an explosive rate. Sooner or later, the lack of energy available will bring back the true definition of the word 'requirement'. Until then, the social and economic barriers to energy conservation may well prove impossible to cross.

One of the reasons for the current world-wide increase in food prices is the use of petro-chemical fertilizers in th agriculture industry, which are made from fossil fuels. I might be a good idea to go back to having our plough pulled by oxen, whose dung would fertilize the fields. Thi is not possible, for several reasons. These methods o agriculture were practical for thousands of years, but the were very intensive in their use of time and labour. Toda the peoples of the world have rising expectations of mor leisure time and more material goods. In addition, ther may already be too many people in the world to grow enough grain this way to feed both the people and the oxen

The natural energy cycle. Most of the water evaporates from the sea, then is precipitated. Fossil fuels are formed by heat and pressure on decomposing vegetation.

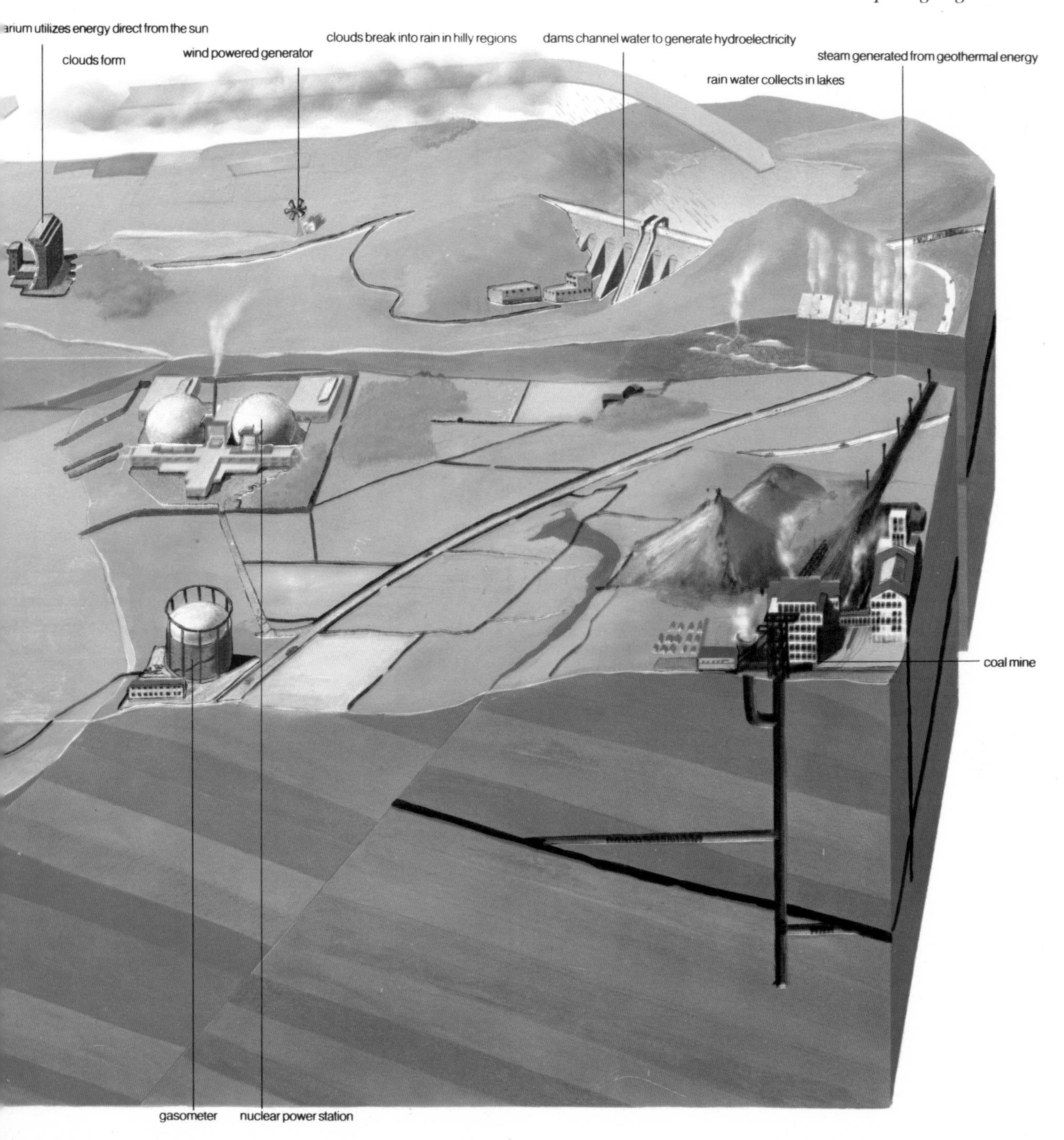

In Britain, many homes are not insulated or heated roperly, so that they are cold and damp much of the year. 'he ubiquitous 'electric fire' (bar heater) is one of the most npleasant as well as inefficient means ever devised of eating a room. On the other hand, Americans on the verage have more rooms and larger rooms than the British, nd think nothing of heating them to 75°F all year round, vhether the rooms are in use or not. Indeed, the American vay of life is the biggest stumbling block for energy conervation, for two reasons: firstly, their wastefulness is self part of their high standard of living, and they will not give it up easily; secondly, it is the American standard of living that the rest of the world wants to emulate.

Economists are not much help. They are usually employed by self-interested groups, and they are not paid to point out that the true cost of doing something is the cost of not choosing an alternative. Thus scientists and economists who have not much social consciousness may say that the 'fast breeder' nuclear reactor is the answer to the energy problem, without taking into consideration the fact that it is impossible to dispose of the poison manufactured by such machines as a by-product. Economists also depend

Below: a pumped storage scheme in North Wales. The upper reservoir is 1000 feet (300 m) above the power station; it produces 360 MW.

too much on their neat mathematical models: a few years ago no industrial economist making a forecast of future costs could have taken into consideration the recent increase in oil prices.

It is possible to conserve a great deal of energy by insulating homes properly, using solar energy to run them, growing more food at home rather than depending entirely on an energy-intensive agriculture industry, making increased use of electronic communication to cut down the necessity of travel, and above all by allowing the input of true energy costs into capitalist economics. All this, however, implies a considerable change in social attitudes; one of the biggest problems is that a politician cannot get elected by telling people the 'bad' news. Governments in industrialized countries are beginning to see the importance of new energy policies, but the people do not want to pay more for their fuel, and there are very powerful pressure groups from the energy industry itself to deal with.

Properly speaking, there is no shortage of energy in the Universe; the problem is that useable energy is being turned into non-useable energy at an increasing rate. In other words, entropy is on the increase, at least so far as planet Earth is concerned. Estimates of energy sources and energy requirements may turn out to be quite wrong; this chapter gives current information based on the assumption that energy use will carry on in its present form.

The requirements

The units of energy are joules, or in a more familiar unit, calories. The energy in food is usually quoted in kilocalories, or Calories. Power stations, however, are usually rated in power units, watts, which do not take into account the time over which the power is used. Energy sources can therefore be quoted in a variety of units, but here the basic power unit of watts (and the multiples kilowatts, kW, and megawatts, MW) will be used.

Humans need only 0.15 kW per man as food for light effort (1 kW = 239 calories/second), but European man used a total of 5 kW per man in 1970, both for food and for other uses, to provide the material standard of life to which he is accustomed. In the underdeveloped countries the figure was about 0.5 kW, while in the USA it was 10 kW. About a quarter of the energy consumption of a developed country is used for room heating and household appliances, a quarter for transport, and the remainder for industrial and agricultural production.

So far most of the energy has been obtained from the combustion of fossil fuel—coal, oil and natural gas. In the

Below: a nuclear power station. The fission process produces radioactive waste material which must be stored for hundreds of years.

Bottom of page: solar cells similar to those which power spacecraft. Experimental devices have efficiencies of 18%, but 12% is more usual.

developed countries, the fraction of the energy which is used to produce electricity was about $\frac{1}{4}$ in 1970 and will probably rise to $\frac{1}{3}$ by 1980. The upshot is that by 1980 the world will require energy at the rate of about 13×10^6 MW, of which 4×10^6 MW will be used for the production of electricity. If the whole world were to be brought up to USA levels of consumption these figures would need to be multiplied by six. (1 MW of power = 1340 horse power; and 1 MW yr of energy is equivalent to one thousand 1 kW fires operating for one year.)

The oil reserves may seem large (430×10^6 MW yr) until the cumulative oil consumption is taken into account: in fact they would be exhausted around the year 2020 even if the annual increase in consumption dropped to 5% from 1980 onwards, and even if progress is made in developing economic methods of extracting oil from tar sands and oil shales. In contrast with oil, coal reserves are large; but they will be needed for the production of plastics and oil as well as power. Provided that coal is not used for many more decades as fuel in electricity generating stations (which at present waste at least 60% of the energy), it will be a useful source of hydrocarbons for the chemical industry for another century. A ton of coal contains about 36 litres of tar and about 14 litres of benzole.

FOSSIL FUELS

Coal

Coal deposits were formed from the remains of vast forests of trees, shrubs and plants which flourished in the hot and humid climates of 250 to 400 million years ago. These flora died and rotted and were buried and consolidated under sediments deposited by encroaching seas. The coal seams so formed lay undisturbed until the coming of man. The Chinese are said to have used coal three thousand years ago, but there is no evidence that other ancient civilizations used it. The Venetian explorer, Marco Polo, records in the account of his thirteenth century travels through Cathay that the natives burned a black stone dug out of the mountains.

Many hundreds of years ago, Europe and the British Isles were extensively covered with forests, but in modern times wood as a source of fuel has become comparatively scarce and therefore expensive. The discovery and exploitation of coal has had an important economic motivation. In the time of Elizabeth I, London began to import coal from mines in the northern part of England; from then until the late Victorian period Britain had no near rival in its production of coal. London's factories and homes produced so much coal smoke that it quickly became the dirtiest city in the world. The poisonous, acrid fumes combined with climatic conditions finally became a health menace which led to the necessity for the Clean Air Act of 1956.

The switch from wood to coal that occurred in the seventeenth century established the foundation of the Industrial Revolution which followed. A coal fire is so dirty compared to a wood fire that a whole new technology had to be developed in many industries in order to deal with it. For example, in the breweries when coal fires were first used to dry the ingredients the resulting beer was undrinkable. The use of coal also encouraged the beginnings of modern mass production: in the glass industry coal fires made possible greater production of plate glass than ever before, but the creation of beautiful objects one at a time by glass blowers became much more difficult in the smoky, poisonous atmosphere.

Parallel with the development of the economics and technology of coal has been the problem of safety in the mines. As mines went deeper and became more complicated, technological progress was made only at the cost of human life. Each advance in the daring of the miners led to disasters against which safeguards were then developed. One early solution to the problem of ventilation was to sink a parallel shaft deeper than the one being worked. Birds were taken into mines because they use oxygen faster than humans, and if the air supply became inadequate the birds would die soon enough to give an early warning. Another problem was the flammable methane gas which is given off by coal seams when they are exposed by digging. In the early days of mining, a man would wrap himself in wet rags and crawl along the floor of a shaft, holding a burning torch on a pole above and ahead of him to ignite the gas, whereupon the mine would be considered safe. Today safety methods are considerably advanced compared to these, but even so disasters taking many lives have occurred in the twentieth century.

Types of coal

Coal is classed according to its characteristics when burnt, its weathering qualities in storage, and its content of volatile materials. The three main types of coal are lignite, bituminous and anthracite.

Lignite is brownish to black in colour, crumbles after exposure to air, and is subject to spontaneous combustion. These characteristics give it very poor storage quality. Bituminous is the most common type of coal, including several lower grades called sub-bituminous. It has good storage qualities. Anthracite, also called hard coal, has a shiny black colour and is sometimes made into costume jewellery. It burns slowly and is highly valued as a domestic heating fuel.

Coke is a substance made from certain types of bituminous coal, sometimes mixed with anthracite. It is almost pure carbon and is used in smelting iron ore. Coke is made by heating coal in an airless chamber almost to its burning point, which causes the volatile components (gas, tar and others) to distil off. These products themselves can be valuable by-products. A great number of organic compounds, for example, can be derived from coal tar and crude benzole. Coal tar alone contains well over 200 chemical compounds.

Far left: a mechanical scraper removing overburden in a Brunswick strip mine. Note the size of the man next to the machine.
Left: a pit head at Ebbw Vale, Wales. The relative compactness of the plant belies the scope and depth of the underground works. The wheels and cables are used to haul the 'cage' up and down.

Modern coal mining methods

Coal is mined by two distinct methods: surface mining and underground operations. The choice of method is dictated by many factors such as the seam thickness, the depth and inclination of the seam, the location of the deposit, surface topography and land value, environmental considerations, economics, and so on.

Surface mining—usually known as opencast or strip mining— is carried out by stripping away the strata overlying the coal seams and then removing the exposed coal. Until comparatively recently it was only feasible to remove a maximum of about 100 feet (30 m) of surface strata (overburden) but the post war years have seen the development of huge excavating machines capable of stripping several hundred feet of overburden. The loading buckets on such machines are capable of scooping up several hundred tons with each operation. To produce one ton of coal it may be necessary to strip as much as 30 tons of overburden, which provides some concept of the size of a modern open pit. The productivity of miners in open pits is very high compared to underground workers and outputs of 50 tons a manshift are recorded. Total mining costs are low by underground standards and may be only one quarter of those in deep mining operations. In Britain less than 10% of national output is opencasted but in the USA nearly half the output is mined this way.

Opencast or strip mining has become an issue among conservationists, particularly in the USA. In some places whole tops of mountains are being cut off. This affects not only the natural beauty but also the drainage of the land, wildlife habitation, and so on. One solution would be to put some of the overburden back and replant it when the surface mine is exhausted, possibly using rubbish collections brought from the cities as additional landfill.

Most of the world's coal is won from underground mines, some of which are 3000 to 4000 feet (900 to 1200 m) deep. In such depths, access to the seams is by vertical shafts equipped with hoisting machinery, but in shallower depths down to 1000 feet (300m) the workings may be connected to the surface by inclined tunnels. Conveyors are usually installed in these surface slopes.

There are two principal methods of underground working: room and pillar, and longwall working. With the former system, once access to the seam has been gained, tunnels (rooms) are driven into the seam in two directions at right angles so dividing the seam into a number of rectangular blocks of coal (pillars) which may or may not be subsequently extracted. Depending on certain practical considerations such as the degree of roof support needed from the pillars and the type of machinery being used, the rooms are nine feet to 24 feet (3 to 7 m) in width and the intervening pillars from 30 feet (9 m) square to 150 feet ×

300 feet (45 × 90 m). Machines have been developed for driving these tunnels which eliminate the need for manual breaking or shovelling of coal. Such machines can cut tunnels in the seam at speeds of up to several inches a minute to produce coal at a rate of up to ten tons a minute. This coal is mechanically taken to the rear of the machine and loaded onto conveyors or wagons for transport to the surface. Room and pillar working is favoured when mining beneath surface buildings or under lakes and seas. Under such circumstances the pillars are left in position to minimize movement of the ground at the surface. When the pillars are left, the term partial extraction is applied to the system.

Longwall working is a total extraction system: all the coal within a specified area is extracted in one operation. Two parallel tunnels are driven into the seam some 150 to 600 feet (45 to 180 m) apart. These tunnels (gate roads) are then joined by a road at right angles, this third road forming the longwall face. Successive strips are then taken off the side of the face road and the coal is deposited on a face conveyor which delivers it to the gate road conveyor and the shaft. As the longwall face moves forward, the roof behind the face is allowed to collapse, the gate roads being correspondingly advanced and supported. Such faces can advance several yards a shift and produce a daily output of 7000 tons. Many longwall cutter operations have been automated, as have the accompanying roof support systems. It is now possible for all operations to be performed by one operator situated well back from the production area. The most advanced systems include a nucleonic probe, an automatic steering device which sends a radioactive low-frequency pulse into the seam above and below the machinery, guiding the cutting operation along the seam and away from the denser surrounding rock.

Drilling for oil

The first producing oil well was drilled in Pennsylvania, in 1859. Since then over two million bore holes have been sunk worldwide. Many of these have failed to find commercial quantities of oil, as opposed to the exceptionally productive few. Drilling is a very expensive business, and a costly gamble too; hence the importance of preliminary geological surveys.

Offshore oil wells are more expensive than those on land, but they are not basically different. Most of the world's oil wells have so far been drilled on land but now, partly because most of the likely land areas have already been explored, drilling at sea is increasingly important. Drilling for oil has been described as analogous to a dentist drilling a tooth with his patient the length of a football field away. This gives an idea of the problem involved in controlling from the surface a drill at the bottom of a well up to 8 km (5 miles) deep.

The drill string and bit

Oil drilling is done by rotating a drilling bit to make a hole. The bit may be a fishtailed steel one for soft ground, but it is usually a rotary bit with hardened teeth. In very hard rock, diamond or tungsten carbide teeth are used and it may take an hour to drill 2.5 cm (1 inch). (In softer rock, however, rates of about 100 metres (or yards) per hour are possible). The bit is fixed to a 'string' of drill pipes which rotate it as it bores the hole. Each length of pipe is normally 9 m (30 ft) long and about 11 cm ($4\frac{1}{2}$ inch) or 14 cm ($5\frac{1}{2}$ inch) in diameter. The pipes are joined by heavy tapered threads. The pipes situated just above the bit are heavier than those in the rest of the string. They are called drill collars and are used to put enough weight on the drill to force it into the ground while keeping the rest of the drill string in tension. The whole of the drill string may weigh several hundred tons and if it were allowed to bear on the drill under compression the string could easily break or jam in the hole. In fact most of the weight of the drill string is taken by the drilling equipment on the surface.

Below: the first oil well. It was dug by 'Colonel' Drake at Titusville, Pennsylvania, in 1859 using ordinary welldigging equipment. Oil was found at 69 feet.

Rotary drilling

The most obvious part of the equipment on the surface is the derrick, looking rather like an electricity pylon and up to 60 m (200 ft) high. Its height is needed to hoist lengths of drill pipe into place, and to stack lengths of several drill pipes screwed together. The drill string is rotated in the well through a rotating table at the base of the derrick, driven at about 120 rev/min by a powerful motor. This rotating table has a central hole, through which a length of square or hexagonal pipe known as a kelly can slide and by which it can be turned. The kelly is the top section of the drill string and drives the rest of the string as it is turned by the rotary table. The drill string consisting of the kelly, pipes and bit is suspended on a hook from the top of the derrick by cables and pulleys. As the bit cuts into the ground, the kelly slides through the hole in the rotary table. When the bit has descended almost the length of the kelly the drill string is wedged in place, the kelly is disconnected, a new length of drill pipe is added to the string, the kelly is reconnected, and drilling begins again.

This operation will have to be carried out over 600 times in drilling a 6000 m (20,000 ft) well. Each time it is done a team of men have to carry out hard and exacting physical work in connecting and disconnecting pipes and wedges and taking new pipe out of the stack. Sheer hard work, as well as highly developed operating skill, is still a most essential part of oil drilling. As drilling continues the drill itself

Below: modern rotary tri-cone jet bits. The bit is rotated under the weight of the drill collar to force the teeth into rock. This makes the cones rotate about their own axis and they crush the rock. The drilling mud is pushed down through the 'jet' to circulate and cool the bit.

Below: a rig, showing the kelly, rotary table and pipes.

becomes blunt, perhaps after only a few hours if it is in hard rock. Then the whole drill string has to be taken out of the hole so that the bit can be removed and a new one put on. This 'round trip' can take up to a day to do. As the drill pipe comes up it is unscrewed in lengths of three, not in single joints, to speed up operations.

During drilling, specially prepared 'mud', a complex colloidal suspension, usually in water, is pumped down the drill pipes through a jet in the bit, and back to the surface in the annular space between the drill pipe and the sides of the hole. The mud circulates through the well quite slowly and cools and lubricates the drill. It also flushes drillings up to the surface, where they are separated from the mud, which is then re-used. In returning to the surface, the mud coats the side of the hole and helps to keep it from caving in. The mud also helps to control any flow of oil or gas from the well. The weight of the column of mud is generally greater than any likely pressure of oil or gas, so that the oil cannot get to the surface until the weight of mud is reduced. In early wells, before mud was used, any oil or gas found under pressure shot at once to the surface, causing a *gusher* which was both difficult to get under control and liable to catch fire.

Another method always used in modern wells to prevent uncontrolled flow is a blow out preventer. This is an arrangement of heavy rubber-tipped pistons that can be hydraulically closed to shut off the well entirely. The blow out preventer is firmly fixed to the top of a steel casing that is inserted into the well and cemented in place as the well goes down. Depending on the tendency of the strata to crumble, and the drilling programme, casing may be continued all or only part of the way down the hole.

When oil is found the first indication is usually from hydrocarbon analysis of the drilling mud returning to the surface. The oil is tested for quality and flow rate and, if this is satisfactory, production tubing is cemented in and a 'Christmas Tree', so called because of a resemblance in shape of the complex of valves and tubing that makes it up, is fixed at the well head.

An alternative to rotary drilling is turbo drilling, where the drill is driven at the bottom of the well by a turbine operated by the drilling mud or (electro drilling) by an electric motor. Rotary drilling, however, is still by far the most usual method.

Offshore drilling

This is being done in many parts of the world, but the North Sea is one of the most active areas for exploration at the present time. It is also the most difficult area so far explored, because of adverse weather conditions and the distance from the coast of most of the fields. Drilling has been going on for gas and oil in the North Sea since the early 1960s, but this has been in comparatively shallow water. At present intensive oil drilling is being carried out in deeper water, under more difficult circumstances. For-

Below: a semisubmersible oil and gas drilling rig, Sedco 135, searching for gas (often found with oil) in the Maui gas field, New Zealand. Gas firing is routine, to test the find.

tunately the whole of the North Sea is shallow compared with the oceans; much of it is between 30 m (100 ft) and 200 m (650 ft) deep. This is typical of the so-called continental shelf areas which make up about 10% of the world's under-sea surface.

Types of marine platforms

To support the drilling rig, ancillary equipment and crew's quarters, some form of floating platform is needed. The first wells were drilled from converted ships, and these are still in use, but a limiting factor is their tendency to drag even the heaviest anchors during rough weather. Fixed or self-contained platforms are used in shallow water, to a depth of about 30 m (100 ft). Another type of rig is the self-elevating (jack-up) platform, which has an operational limit of 90 m (300 ft), or so, because of bending stress in the leg supports. They can be towed into position and the legs jacked down until they stand on the sea bottom and then further jacked until the platform is well above the sea surface, clear of the heaviest waves. The most recent development in offshore drilling has been the use of semisubmersible rigs. These have several large hulls with long legs holding a platform above them, and the hulls are ballasted so as to sink about 20 m (65 ft) below the surface of the water. As with jack-up rigs, the platform is still well above he water and clear of the waves. The rig may be held in place by multiple anchors or it may be dynamically positioned. In this method, multiple propulsion units on he rig respond to signals from a beacon on the sea bottom and keep the rig exactly in position in relation to the beacon, even in the worst weather.

At the hull level a semisubmersible may be about 60 m (200 ft) wide by 76 m (250 ft) long, and its operating draught will be 18 m (60 ft) to 27 m (90 ft). One rig could cost more than thirty million dollars. The biggest semisubmersible in the North Sea at the end of 1973 could drill to a depth of 10,000 m (33,000 ft) in up to 300 m (1000 ft) of water. It could survive in winds of up to 220 km/h (136 mile/h) and in waves of up to 26 m (85 ft). Even larger semisubmersibles are currently being built.

Producing wells

After exploration drill rigs have been used to find oil they are moved on to other areas for further exploration. In order to drill producing wells, production platforms are installed. These enormous steel or concrete platforms stand on the sea bottom and, by angling the hole using a technique known as directional drilling, up to 30 producing wells can be drilled from each platform. Oil is treated on the platform to remove gas and water and is brought ashore by pipeline or by tanker. Plans are also being made to drill producing wells and take oil from them without a production platform, working on the sea floor from wellhead cellars serviced by pressure vessels from the surface. A system of this kind has been used to drill and service a well in the Gulf of Mexico in 114 m (375 ft) of water, and a North Sea trial was being planned for the late 1970s. This undersea technique could show great savings compared to drilling only from platforms but it will probably supplement rather than replace the production platform.

Oil refining

Crude oil, as it comes from the oil well, will always burn. But except for some applications, such as operating a pump engine on a crude oil pipeline, it has to be refined to make it a satisfactory fuel. Also, it would be impossibly wasteful to burn crude oil, so losing many of its valuable constituents, such as chemicals feedstock and lubricating oil. The feedstock is the part of the crude that goes to the chemical industry for making everything from fertilizer to gramophone records.

Oil refining is therefore always necessary, and it does two main things to crude oil. It divides the crude into fractions of differing volatility, each with its own special application, and it alters some of these fractions physically or chemically either to make them more suitable for their own applications or to make it possible to use them for other applications where the demand is greater. For instance, crude oil can be divided into a gaseous fraction, and liquid fractions that form the basis of petrol (gasoline), paraffin (kerosene), diesel oil, fuel oil and, perhaps, bitumen or lubricating oil. For petrol (gasoline), however, the fraction produced by separation is not good enough in quality for a car engine to run on it and must be improved by refining processes. The crude oil may also not yield enough of the fraction in the right range to meet the demand, and other fractions of the yield may have to be altered so that they can be incorporated.

Refining, producing over 3000 million tons of products a year worldwide, is therefore a complex business. It is made more complex by the great variation of crude oil composition according to the area it comes from. Refineries must be able to produce the required balance of products from a range of crude oils and they must also be able to change the balance as demand changes. The change in demand may be relatively long term, as when vaporizing oil for tractors was replaced by diesel oil, or it may be seasonal—for example, the demand for heating oil increases in the winter. Seasonal demands can to some extent be met by storage in the off-season, but refineries must arrange to make more of the product in demand at the expense of products less in demand. The scheduling needed to do this at maximum efficiency is so complicated that computers must be used to plan it.

Most refineries are built at deep water ports because they offer a number of advantages: large tankers can readily discharge crude oil to the refinery; the refined products can be transported cheaply by water to the markets; and ample cooling water is available for refining processes. A large refinery will require crude oil storage tank capacity of several hundred thousand tons, with some tanks taking up to 100,000 tons each, to accept full cargoes from modern large tankers. Such a refinery may handle 20 million tons of

Below: a crude oil distillation unit with a capacity of 160,000 barrels a day, In the beginning of the refining process, the crude is split into fractions, depending on their boiling points.

Below: a large part of the area of a refinery consists of tanks for storing various product, and as much as 1000 miles of pipeline.

Next page: refinery processes. If a natural crude doesn't have enough of a particular fraction, conversion methods such as cracking are used.

crude oil in a year, and will represent an enormous investment. As well as the plant for processing the crude oil, the refinery will need auxiliary services such as power and steam supply, and maintenance and laboratory facilities.

Even large refineries, however, do not need much manpower to operate them and an operating shift could consist of only about 25 people. The largest use of manpower is for maintenance teams, to keep the refining plant in service and to maintain the complex automatic equipment and controls that keep a modern refinery in operation. Automatic control of processes is not primarily intended to save manpower. It has developed to its present highly complex form—where just one refinery unit may have up to 200 automatic controls on it—because automation gives more efficient and safer operation.

Gas production

The modern gas industry has its origins in the late 18th century development of gas lighting, but the use of gas for heating and cooking did not become widespread until the second half of the nineteenth century.

The first company to manufacture and supply coal gas on a commercial basis was Boulton and Watt, of Birmingham, England. This company, founded by Matthew Boulton and James Watt, employed William Murdock as a steam engine erector, and it was the experimental work of Murdock which led to their early involvement with the gas industry.

Boulton and Watt gave up their interests in gas making in 1814, and the development of the industry was then continued by the Gas Light and Coke Company in London. Their chief engineer was Samuel Clegg, who had studied under John Dalton and been an apprentice at Boulton and Watt, and it was Clegg's expertise which helped the Gas Light and Coke Company to develop the system of distributing gas by pipeline from a central gasworks to the customers.

The American gas industry began in Baltimore, Maryland, in 1816 when the city council authorized Rembrandt Peale to manufacture gas and distribute it by pipes laid under the streets. The modern use of natural, rather than manufactured, gas began in the middle of the nineteenth century, when the Fredonia Gas Light and Water Works Company was formed in 1858, and there are now over 100,000 natural gas wells in the United States alone. Extensive natural gas deposits have also been found in North Africa, Russia, Australia and off the northern European coast.

The three major types of gas are coal gas, oil gas, and natural gas, which is now the most important of the three. Liquefied petroleum gases such as butane and propane are also distributed in some areas.

Coal gas

Coal gas is made by the carbonization of coal, a process which involves heating the coal in the absence of air to drive off the gas, and which also yields several useful by-products. The most important constituent elements of coal are carbon, hydrogen, oxygen, nitrogen and sulphur, and it is the hydrogen and carbon which form the basis of coal gas.

There are many factors affecting the exact composition of

coal gas, including the type of coal used, the manufacturing temperature, and the type of retort in which the coal is carbonized. The crude gas obtained from the coal is passed through several purification stages to remove unwanted constituents such as ammonia, hydrogen sulphide, hydrogen cyanide, tars, and various hydrocarbons, before it is suitable for use.

Oil gas

The first applications of petroleum oil for the manufacture of gas were those in which crude oil was thermally 'cracked' with steam at atmospheric pressure and at temperatures of between 1000 and 1100°C (1832°F and 2012°F).

Subsequently the range of process materials was extended

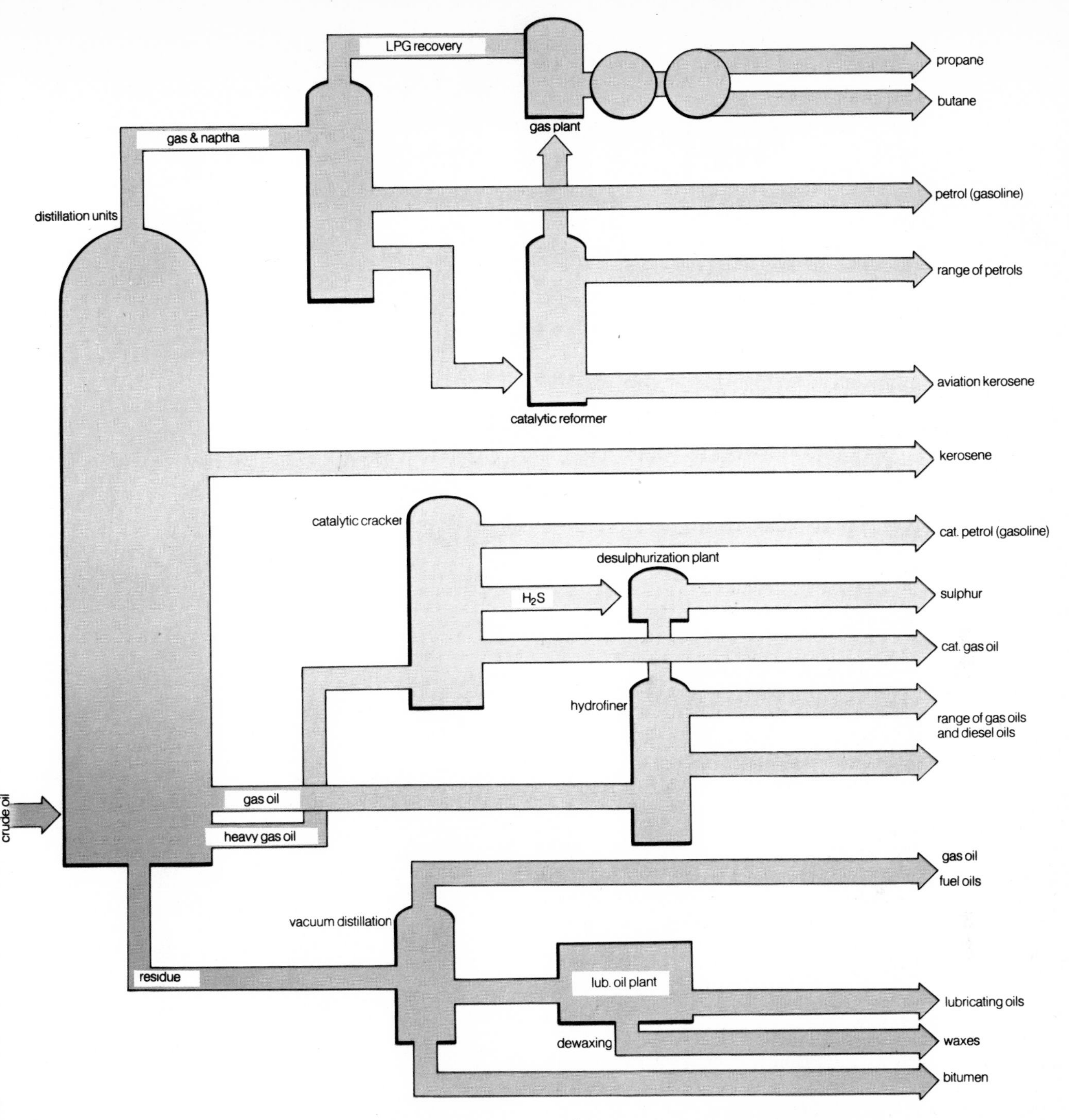

) include most liquid hydrocarbon 'feedstocks', but this urely thermal cracking process was superseded by a more fficient cyclic catalytic process and eventually a whole ange of cyclic and continuous processes called reformers vere devised.

Oil gas processes lent themselves more readily to full nechanization and automatic control. The purification of il gas compared with coal gas is simplified owing to the bsence of ammonia and hydrocyanic acid, and the overall esult of changing to oil gas was a relatively lower capital ost and lower operating labour costs. The introduction of il gas was the first large scale revolution in gas making echnology since the first commercial applications of coal as manufacture.

The objects of catalytic processes are twofold: firstly to roduce gas from hydrocarbon feedstocks with a reduced ield of by-products; and secondly to obtain higher gas ields compared with those attainable by unaided thermal racking in steam.

Large scale cyclic and continuous reformers do not differ greatly in terms of capital cost and thermal efficiency, but the pressures achieved in continuous reformers are very much higher, resulting in a greater gas yield. This represents the greatest single advantage of the continuous process.

Natural gas

Natural gas, of which the main constituent is methane, is usually found in the same type of geological strata as oil, and is often found in association with oil deposits, the oil being driven to the surface by the pressure of the gas. In the early days of oil production the gas was regarded as a nuisance and was merely piped clear of the oil lines and burnt away. The Chinese used natural gas for lighting, distributing it in bamboo pipes, as long ago as the first millennium BC.

Natural gas became the major source of gas in the USA in the 1930s, following the discovery of extensive deposits and improvements in pipeline technology. In Europe, one of the

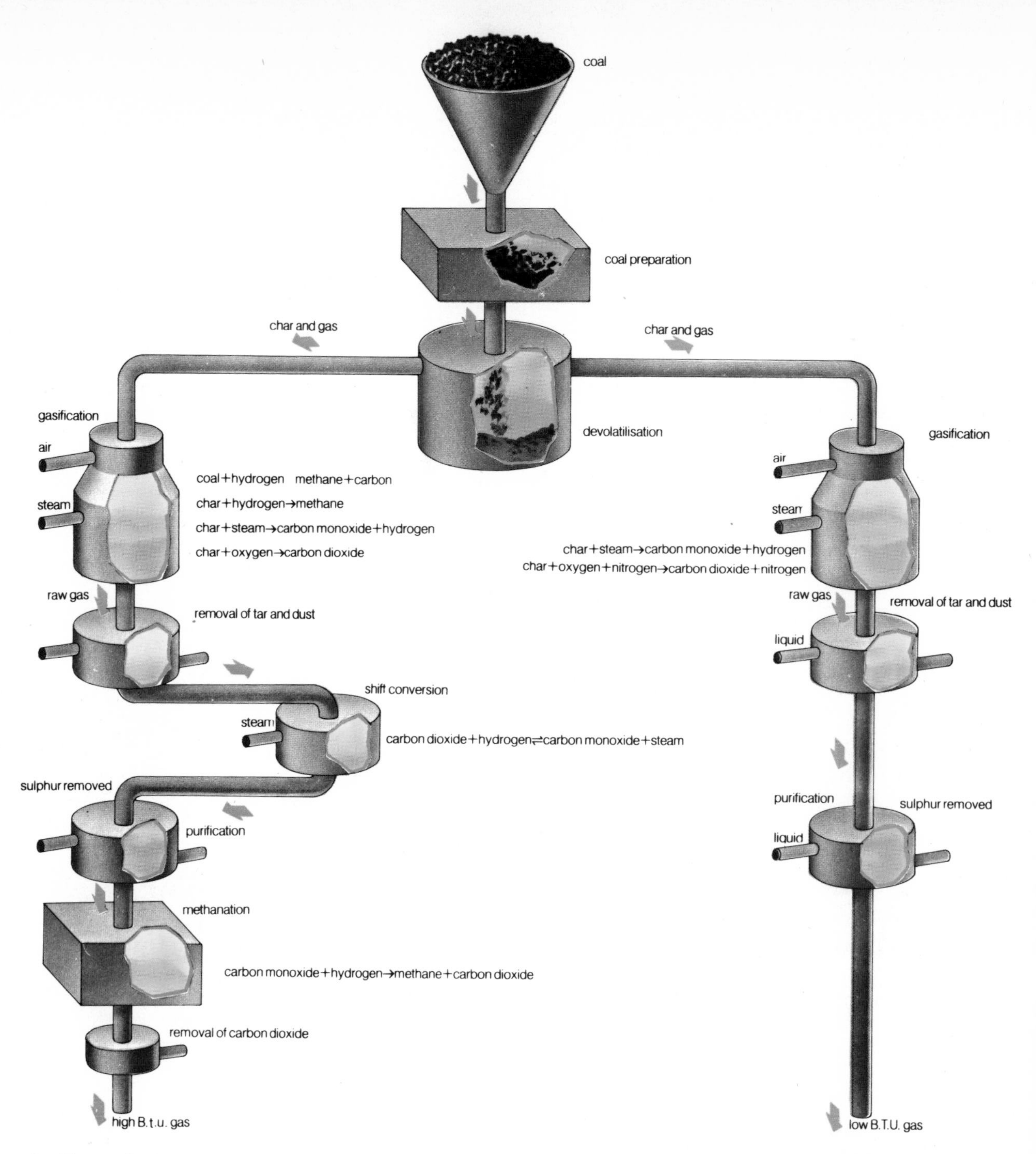

most significant finds was made at Groningen, northern Holland, which strengthened the opinions of many geologists who believed that there was gas, and possibly oil, in vast quantities under the North Sea between Britain and Holland.

The first commercial find in the North Sea was in 1965. Exploration and discovery made steady progress, and it was soon being piped ashore for use in Britain. There was a problem, however, in that the calorific value of natural gas is about twice that of manufactured gas. The calorific value of gas is the amount of heat available from a given volume, and all the 40 million gas appliances in the country were designed to burn this gas and could not burn natural gas safely or efficiently.

There were two solutions to this problem: to use natural gas as a feedstock for the manufacture of gas by an adaptation of the oil gasification methods available (reforming); or, and this was the solution chosen, to convert the whole gas supply system of the country to handle natural gas. This decision was taken mainly for economic reasons, as i was estimated that the cost of conversion would be £40 million spread over 10 years, but over a period of 30 year the total cost of reforming the gas would be £1,400 millio more.

The North Sea gas is brought by pipeline from the off shore drilling rigs to coastal reception terminals, an distributed by a 2000 mile (3219 km) long high pressur pipeline network, operating at up to 1000 psi (68.95 bar). The gas is taken from the main grid and distributed locall at lower pressures. By 1974 90% of the conversion worl had been completed, and because of the increased calorifi value of the gas, existing pipelines were automaticall doubled in energy-carrying capacity.

Natural gases vary in content, and some contain amount of heavier hydrocarbons which are removed to produc liquefied petroleum gases, and other gases must be treate to remove unwanted sulphur compounds and carbo dioxide. A typical treated natural gas contains over 80%

ft: two coal gas methods, ·h of which begin with ·olitization (carboniza- n) of the coal, which oduces 'char' (carbon) d a mixture of gases.

Below: a natural gas plant in Scotland. The gas is liquified for storage; the capacity of the plant is 1000 million cubic feet.

Bottom: the interior of a side-firing reformer which produces town gas from other gases. It can also make synthesis gas.

methane together with ethane and about 1% nitrogen. Natural gas has no appreciable smell of its own, so to prevent accidents a small amount of a smelly gas is added.

Other processes

There are several other important gas making processes. The water gas process involves blowing first steam then air through a bed of heated coke. The gas produced is often called 'blue water gas' because it burns with a characteristic blue flame. Carburetted water gas is produced by using the hot gases from the water gas generator to crack oil, producing an oil gas which is mixed with the water gas to enrich it.

The Lurgi process, introduced in Germany in 1945, uses steam and oxygen under pressure to make gas from lignite or low grade coals. Liquefied petroleum gases, chiefly butane and propane, are by-products of oil refinery processes and natural gas treatment. Coke oven gas is a by-product of industrial coke making, and blast furnace gas is produced during the iron smelting process and is used at the works for steam raising, power generation, and pre-heating the blast air for the furnace.

Fossil fuels represent the accumulation of 400 million years of solar energy transformed by photosynthesis in plants. On any reasonable time scale they must be regarded as non-renewable resources, and the end of man's brief fossil fuel period of 2000 years is in sight. What are the alternatives?

WATER POWER

Evaporation by the sun and rainfall on high ground represents the largest renewable concentration of solar energy. The power of water is harnessed by allowing it to fall under gravity through turbines which drive electric generators, and consequently this source of energy is referred to as hydroelectric power. The potential world capacity of hydroelectric power is about 2.9×10^6 MW but only 7% is being used. Unfortunately many of the unused sources are far from centres of population and industry, and transmission costs, which are very high, cannot be ignored. Furthermore, such schemes may affect the environment, for example by altering the flow of rivers and causing silting up. Therefore, although useful for many under-developed regions, hydroelectric schemes can be no answer to the needs of the developed world. (See also *Generation of electricity* in the chapter on electromagnetics.)

Below: falling water possesses kinetic energy which can be used to turn turbine blades and generate electricity, as in this hydroelectric plant in Sogsvirkjun, Iceland. Opposite page: a geothermal power station at Taupo, New Zealand. Steam from the Earth's internal heat is tapped and fed directly to turbines.

GEOTHERMAL POWE

Steam is available from hot springs in volcanic regions a the total installed generating capacity is about 1200 M Assuming that about 1% of the potential energy availab can be tapped, and converted to electricity with an efficien of 25%, the potential yield is estimated to be 3×10^6 M yr. If withdrawn over a 50 year period, this source wou provide about 60,000 MW of power. Highly significa though it is to a country like Iceland which has no fos fuel, this source is irrelevant in global terms.

The heat energy in these volcanic regions is thought arise from a near-surface concentration of a more wide dispersed heat source of radioactively decaying mater about 20 miles (30 km) below the surface. Deep drilli projects might lead to this source of geothermal heat bei tapped in regions where the hot rock is not too dee although the power required to pump a heat transfer flu through fissured rock would be considerable. The he available from liquid granite magma at 900°C, as it coo and crystallizes at 500°C, is about 60,000 MW per cub kilometre. At 20% conversion efficiency, which is opti mistic, a cubic kilometre of magma could support a 10(MW station for only 12 years. Geothermal energy unlikely to be the answer to the energy crisis.

WIND POWER

Below; a windpump on a hill in South Africa, with storage tank for windless days.
Bottom: the machinery at Pitstone Mill, England, dates from before 1627.

At the present time it must be said that efforts to generate electricity by wind power on anything but a very small scale have failed. What the future holds remains to be seen, and as governments throughout the world search for sources of power alternative to fossil fuels and atomic power, there can be little doubt that large sums of money will be spent upon experimental work.

The small battery charging wind generator can be quite successful if it is well designed and maintained, mounted on a sufficiently high pole in a region of steady wind, and is close enough to the point where the electricity is to be used. It is not always appreciated that generation is usually at a low voltage, so that transmission over anything but a short distance results in high losses or high capital cost, and mounting a wind generator above a house top is unsatisfactory because of the eddies caused by the building.

The normal system is to charge low voltage batteries or run direct if the wind is steady and its speed high enough. The advent of an ample supply of electricity to most dwellings in Western Europe and North America and the relative cheapness of the small engine-driven generator which produces AC at the correct voltage and periodicity for domestic appliances has removed much of the incentive to install small wind generators, except in isolated places.

It must be appreciated that the specific volume of air is low, as is the average wind velocity. On the other hand, a strong gale can do great damage to a high and lightly built structure, and may well destroy it. During World War II considerable experimental work was carried out in Denmark, where the winds are strong and there was an acute fuel shortage, but firms which had been making 5 to 25 kW wind generators did not continue doing so after the war. It is generally accepted that wind power for larger units would best be developed by using propeller-type windwheels driving induction generators and feeding into a network. The wind generator could not provide firm power, and the cost evaluation must be solely upon the basis of the number of units generated. In other words, it must be a fuel saver only, and the economics calculated as such.

Several experimental windmill generators have been built since World War II, but none have been successful. One was built in the USA at a place called Grandpa's Knob in 1947; there were two smaller ones built in the UK in 1954. One of these was at Enfield, and had a hollow two-bladed propeller with orifices at the tips. It sucked air up a tubular tower as it rotated, and this drove an air turbine coupled to a synchronous generator.

The most ambitious design, from Germany in 1947, has not actually been built. If it is accepted as a possibility, it is necessary to imagine a propeller 425 feet (130 m) in diameter, together with a generator, mounted on a tower 820 feet (259 m) high, and capable of rotating 360°. This tower would be nearly as tall as the Eiffel Tower, but would have to be capable of withstanding gale force winds. 47 such monsters would be required to replace one 660 MW steam generator, and could only do so if the wind speed were over about 25 mile/h (40 km/h), described as a strong breeze (Wind Force 6 on the Beaufort scale).

TIDAL POWER

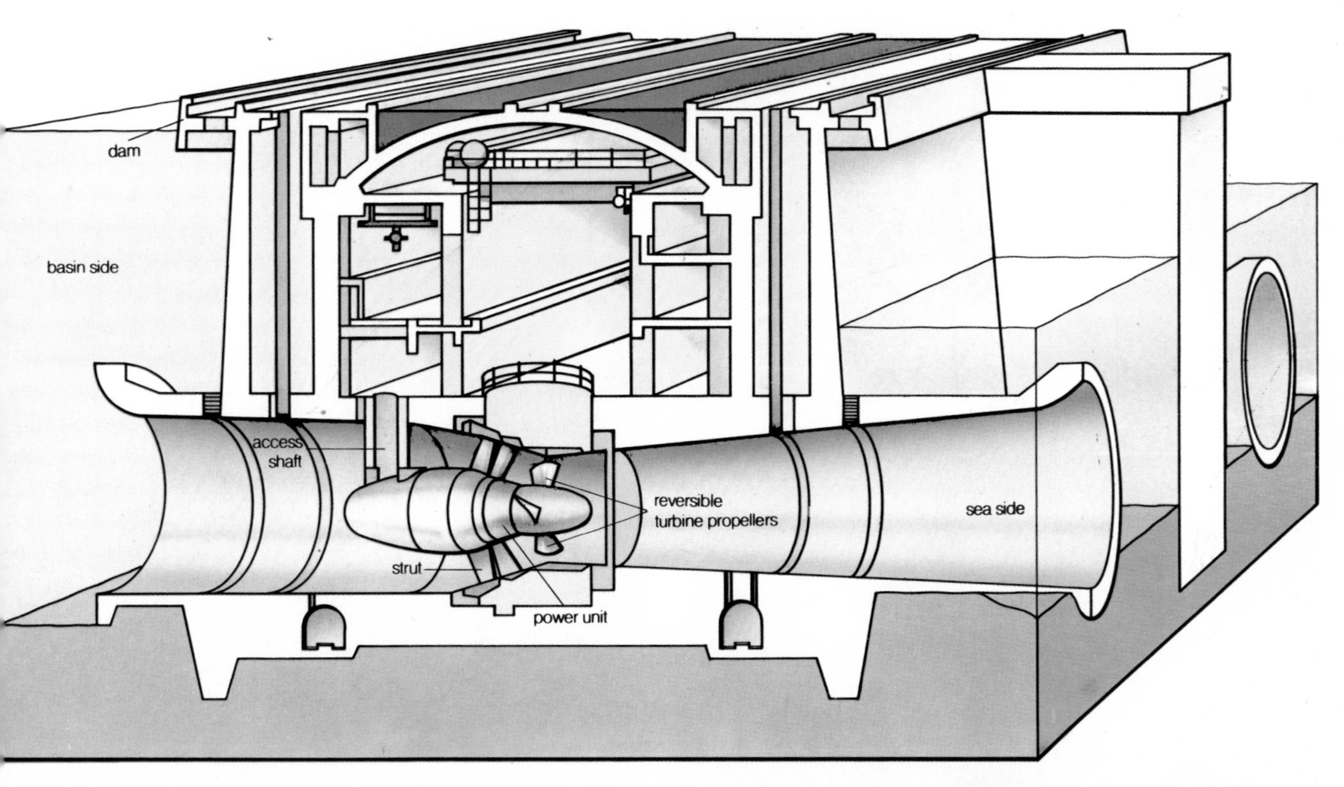

Although tides are an unavoidable inconvenience to shipping, they have also been used as an energy source, first for mills and more recently for the generation of electricity. The world's only tidal power station, in a barrage on the river Rance in France, avoids the problem of a limited period of operation by using surplus electricity during off-peak hours, as well as the tide, to store water above the barrage. The principal drawback of tidal power production is that the head of water is about ten times less than that of a normal hydroelectric power station. Since the power generated depends on the head multiplied by the rate of flow of water, a very high rate of flow, and hence large and expensive turbines, is necessary. Even so, the output is considerably less than that of a conventional power station, and tidal power can at best provide only a small though possibly significant contribution to a nation's power requirements.

Tide mills

The principle of a barrier across a river mouth to store tidal water has been used for centuries for tide mills. The incoming tide is allowed to flood a reservoir or creek above the mill, and sluice gates are closed (sometimes automatically, using the outgoing water flow) just after high water. A few hours later, when the tide had gone down, the pent-up water would be released to drive the mill wheel. This produced greater power than if the tide had flowed out at its normal rate. Some wheels were adjustable, to compensate for the varying water level in the reservoir.

Salt water does not penetrate very far up a river estuary when the tide comes in, but the rising tide does reverse the flow of river water for a considerable distance upstream.

Top of page: a cross-sectional drawing of the tidal barrage on the French river Rance. The turbine assembly with its sealed generator is called a bulb set, because of its shape. There are 38 of them, each producing 9MW with a tidal head of 15 to 30 feet (5 to 10m).
Above: between 1581 and 1822 there was a large tidally operated wheel in one of the arches of London Bridge, where the rapid flow of water through the arch produced a useful head of water. It was used to pump water from the Thames to the centre of London, and was raised or lowered with the water level. This illustration is from 1749.

SOLAR RADIATION

Below left: modern laboratories use parabolic reflectors to produce high temperatures.
Below: the French solar furnace, also seen on the next page.

The earth receives enormous amounts of radiant energy from the Sun, which directly or indirectly sustains all living things. Anxiety about dwindling supplies of fossil fuels and about the problems that accompany nuclear power has led to rapidly growing interest in the possible ways of harnessing solar energy in ways useful to man.

Energy available

The sun radiates energy at a virtually constant rate of 3.8×10^{20} MW (380 million million million megawatts) by nuclear fusion processes in its interior, which cause it to lose mass at a rate of some 4 million tons per second. The orbit of the Earth round the Sun is slightly eccentric, so that the solar intensity just outside the atmosphere varies slightly from 1399 MW per square kilometre at the closest (January 3) to 1309 MW per square kilometre at the furthest (July 4). The energy distribution in sunlight as seen outside the Earth's atmosphere approximates fairly closely to a black body radiator at 6000°K, but this spectrum is modified as the sunlight passes through the Earth's atmosphere. (A 'black body' is one which absorbs and radiates energy perfectly.) The ultra-violet component is cut down by absorption in the ozone layer at about 30 miles (50 km) altitude, and parts of the infra-red are absorbed by water vapour and carbon dioxide. Moreover, some light is scattered out of the beam by dust particles and molecules of the air. Most of this scattered light reaches the ground in diffuse form, however, and since short wavelength light is the most strongly scattered, the sky generally appears blue.

The extent of all these effects depends on solar altitude an local atmospheric conditions. The insolation, or sola intensity, at any point on the Earth's surface therefor varies with the season and the time of day in a regular way and irregularly with cloud cover. The maximum intensity of about 1000 MW/km², occurs when the sun is overhea in a clear sky, but average figures are lower than this Highest annual insolation occurs in the tropical deser areas (near the Equator, insolation is somewhat reduced b high humidity).

The solar energy that falls on the Earth's surface heat the surface of land and water, and evaporates water fron rivers and oceans. Only a small proportion (less than 2% is converted to the mechanical energy of wind and wave and into stored chemical energy in plants by the process o photosynthesis. Solar energy as such is not at presen harnessed by man in significant quantities. Although th energy falling on the Earth in a fortnight is equivalent t the world's total initial stock of coal, oil and gas, large area of collectors are required to collect significant amounts o solar energy, and the price of conventional fuels is only jus rising to the point at which the capital expenditure on suc collectors would be justified.

Solar water heating

When solar energy falls on a black object, which absorb the light, the object is warmed, and in this way radian energy can be readily transformed into low temperatur heat. One of the simplest, as well as the most widespread

Left: the solar furnace at Odeillo in the Pyrenees. 9000 mirrors form a north-facing parabolic reflecting surface with the furnace room (see previous page) at its focus. 11,000 flat mirrors on the hillside opposite can be moved to direct the Sun's rays on to the curved reflector; in minutes, the pollution-free furnace can reach 33000°C (6000°F). Below: solar heat distilling brackish water in Australia. The heat evaporates the water, which then condenses and is collected on the glass plates.

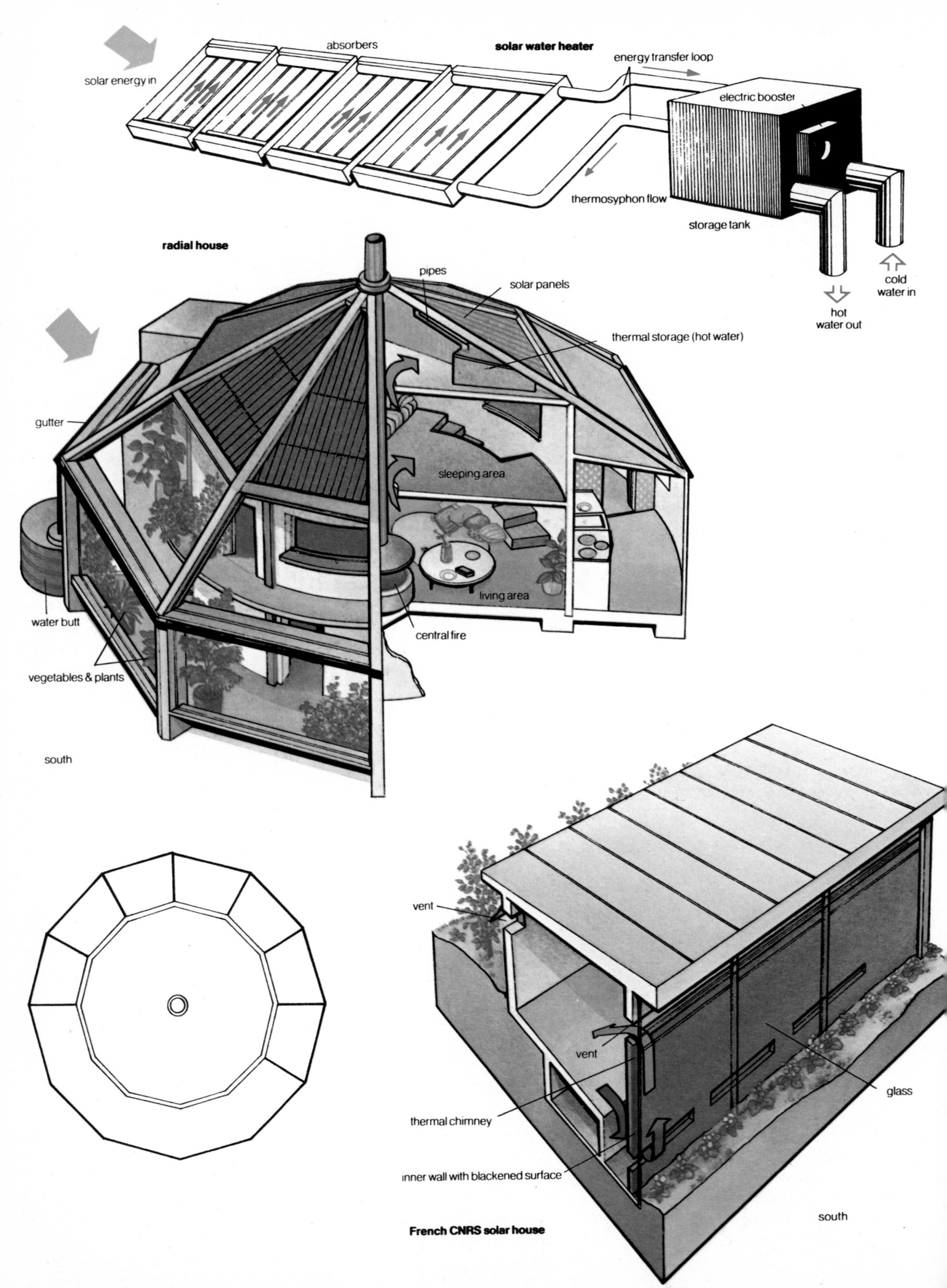
solar water heater
solar energy in
absorbers
energy transfer loop
electric booster
thermosyphon flow
storage tank
cold water in
hot water out
radial house
pipes
solar panels
thermal storage (hot water)
gutter
sleeping area
living area
water butt
central fire
vegetables & plants
south
vent
vent
glass
thermal chimney
inner wall with blackened surface
south
French CNRS solar house

Left: various solar heating systems for houses. The radial house, designed by Girardet, uses panels on the roof. The French CNRS house has a 'thermal chimney'.

applications of solar heat is the provision of domestic hot water, using solar flat plate collectors. Several million of these are in use today, mainly in Japan, Israel and Australia. The units generally consist of a metal plate with a blackened surface, through or over which water flows in pipes or corrugations. The plate is insulated behind to prevent heat loss by conduction, and in front of it there is an air gap of a few centimetres and then one or two glass cover plates, which help to prevent convective heat loss. The collector is placed facing south (in the northern hemisphere) or north (in the southern hemisphere) at a tilt angle which is usually equal to the angle of latitude, so the collector surface is perpendicular to the average direction of the Sun's rays.

Solar radiation (both direct and diffuse) passes through the glass cover plate and warms the metal surface, which in turn warms the water flowing through the pipes. The water is circulated from the hot water storage tank, either by convection or by a small pump. With appropriate thermostat controls to switch the system off when the insolation is too weak to make a useful contribution, solar water heaters can readily provide a reliable domestic supply of hot water in consistently sunny areas of the world. A conventional booster heater is, however, generally fitted in the tank to provide heat during prolonged cloudy spells.

The average thermal efficiency—the ratio of heat falling on them to useful heat extracted—of these collectors is generally in the region of 45 to 65%. The main heat losses occur from the front surface of the collector itself, provided all other parts of the system are well lagged. The glass cover assists in minimizing these losses, since glass is transparent to visible radiation but opaque to infra-red. Thus, solar radiation passes through it but the long wave thermal radiation from the warm collector plate cannot pass outward through the glass. The system can be made more efficient by means of a vacuum between the glass and the metal plate and by using suitable spectrally selective surfaces on the metal. These are surfaces with properties opposite to those of glass. They absorb visible radiation efficiently, so they are black to sunlight, but they do not appreciably absorb longer wavelength infra-red radiation. Consequently they do not emit the latter wavelengths either and so cannot radiate heat, and therefore a spectrally selective superblack body reaches a higher temperature in sunlight than does a normal uniformly black body. Solar water heaters incorporating an absorbing surface of superblack anodized galvanized steel are commercially available. They absorb 93% of visible light but emit only 10% of their infra-red, and this produces an improvement in average performance of several per cent.

Solar houses

Part of the heating and cooling requirements of a building can be provided by the solar energy falling on the roof or walls. Many designs have been proposed, and there are several hundred experimental solar buildings in existence, most of them in the United States, where the study of solar applications has been much advanced by the passing of the Solar Heating and Cooling Demonstration Act in September 1974. This Act provided, among other things, $60 million over five years for the design and erection of solar heated and cooled buildings, of all sizes from small family houses to large commercial buildings.

Solar heating occurs to some extent in all buildings, owing to sunlight warming the material or passing to the inside through windows. All well-insulated buildings carefully designed to have a natural balance at a comfortable temperature are really in part solar heated, though they are not commonly thought of as solar buildings. There is a solar heated annexe of this type in St. George's School, Wallasey, Cheshire, in the UK. This building has a south-facing glass wall combined with unusually high standards of insulation in the other three walls. It is heated almost entirely by solar energy, with some contribution from the heat generated by the lighting system and the occupants. Problems with ensuring an adequate supply of fresh air without losing too much heat have not, however, been solved.

The *Centre National de la Recherche Scientifique* of France has developed an interesting solar-driven air conditioning system which can provide warm or cool air for buildings as required. Thirty-six houses incorporating the CNRS system have been built in the South of France with funding from the French government to evaluate the design. Each house has a black-painted south-facing wall, capable of absorbing a great deal of heat, with an outer glass wall. Solar radiation is transmitted through the glass and warms this wall, creating a rising column of warmed air between the wall and glass. By adjustment of vents, the warmed air can be circulated to the interior of the house to heat it, or fresh cool air can be drawn through the building in summer. As the wall retains heat for several hours, the system continues to work after dark and during overcast portions of the day.

It is also possible to use solar water heaters to provide space heating and cooling in a building. Larger areas of collectors than needed simply to provide hot water are required, and the solar warmed water is stored in large tanks for circulation in the central heating system. In climates where summertime cooling is required, and night skies are generally clear, the same collectors can be used to chill the water very effectively during the night by pumping it through the collectors, which radiate heat to the sky. The chilled water can then, through an appropriate heat exchange system, be used to cool the house the next day.

A similar system, which uses a shallow reservoir of water on the roof of the building, has been developed in California. When winter heating is required, the reservoir is covered during the day by a black sheet which absorbs solar energy and warms the water beneath. Thermal insulation is rolled over this at night to prevent upward heat loss to the sky, so that the warm water warms the house beneath. During the summer, this insulation, which is white, is placed over the pond during the day to prevent the water from being warmed by the sun, but it is rolled back at night and the water is chilled by radiation loss and evaporation, and the house is subsequently cooled as interior heat is transferred to the cool reservoir.

In all such buildings, the initial costs tend to be rather higher than those of conventional buildings, and this is offset only by several years of fuel savings. As fuel prices rise, however, the economics of solar conditioned buildings are becoming steadily more attractive.

Other thermal uses

Solar distillation of brackish or saline water is carried out quite successfully on a small or medium scale in several countries, and some solar water or air heaters are in use in Australia and Russia for crop drying and timber curing. To obtain higher temperatures, it is necessary to focus sunlight by means of lenses or curved mirrors. Only direct sunlight, not diffuse daylight, can be focused in this way. In consistently sunny parts of the world, there is some interest in the use of focused collectors to boil water to raise steam. The steam could then be used to generate electricity by conventional means, or to provide mechanical power for applications such as pumping irrigation water.

Making electricity

The necessity for continuous electric power generation on space satellites led to the development of the solar cell in the Bell Telephone Laboratories, USA, in the 1950s. These devices, which are generally made from thin slices of highly pure single crystal silicon, produce electric power from radiant energy. Wavelengths in the range 400—1100 nm (nanometres) are the most effective, and about half the solar spectrum falls in this range. The actual conversion efficiency of silicon cells is, however, considerably less than 50%, owing to various internal losses. The best cells available today convert sunlight to electric power with an efficiency of about 18% (the remainder of the energy is degraded to heat in the cell).

These cells contain added dopants (usually boron and arsenic) in small amounts to create in the crystal slice a junction between an *n*-type and a *p*-type semiconductor region. This creates a gradient of electric potential within the crystal, and when light falls on the crystal and excites electrons in it, creating electron-hole pairs, the junction separates the electrons and the holes, and a DC current flows. These cells have no storage capacity, and for terrestrial applications they are used in conjunction with electrical storage batteries. They are at present much too expensive for widespread use on Earth, though new manufacturing methods being developed in the United States may greatly reduce their cost.

Even if the cost were low, one disadvantage would still be the low efficiency. To operate a 500 W electric fire would require 2½ square metres (12 square feet) of cells even with maximum sunlight shining directly on them. Unless a large scale method of energy storage or transmission of electricity is employed, this power would have to be used fairly close to its source of production, where it is probably not needed anyway. One possible solution to this problem in the future may be to equip a satellite with large arrays of solar panels generating power which would be transmitted to Earth by means of microwaves. These could be picked up by receiving stations consisting of arrays of wires even on cloudy days at fairly high latitudes.

NUCLEAR REACTORS

The discovery of radioactivity in 1896 revealed that the elements thorium and uranium release energy spontaneously. The release accompanies a series of radioactive transformations in which atoms emit particles or rays and change their chemical identity. The rate of energy release is too slow to be of much practical use and it seemed that nothing could be done to hasten it.

A breakthrough occurred in 1919, when Rutherford discovered that alpha-rays could shatter the atomic nucleus. Further research led to the discovery of the neutron in 1932 and of the fission of uranium in 1939. That year it became clear that a nuclear chain reaction could probably be set up, using uranium, and that this might be the means, not only of releasing vast amounts of energy, but also of producing a new element, plutonium. There was the possibility that an atomic bomb could be developed and it was to produce plutonium for such a bomb that the Hanford Works was built beside the Columbia River in the USA. Here, the world's first industrial-scale nuclear reactor for the production of plutonium commenced operation in 1944. The heat of the nuclear reaction was carried away by river water. The next step was to develop reactors whose heat could be converted into useful power. This meant higher operating temperatures.

In the USA, the first objective was submarine propulsion, the USS *Nautilus* commencing sea trials in 1955. In Britain, Calder Hall was built, for the dual purposes of plutonium production and electricity generation, and a programme of nuclear power was put before Parliament. In the USSR, an atomic power plant had commenced operation at Obninsk in 1954. Twenty years later, nearly 250 sea-going nuclear propulsion plants and about 120 full-scale industrial nuclear power reactors were in operation by 16 nations.

When a nuclear reactor is operating, neutrons are emitted and absorbed by the fuel, while heat is released in it. As the nuclear fuel 'burns', its nature gradually changes.

The neutron is a nuclear particle which has no electric charge. It can penetrate matter, only occasionally colliding with an atomic nucleus, rather as one might walk blindfold in an orchard, occasionally bumping into a tree trunk. When a collision occurs, the neutron either bounces off in a new direction, or is captured, forming a compound nucleus.

Thermal reactor

Generally speaking, the chances that a neutron will interact with a nucleus are much higher when the neutron's velocity is low than when it is moving fast. For this reason, in the so-called thermal reactors, a moderator is used to slow down the neutrons emitted by the fuel to the velocities of thermal agitation. The slowing down occurs because the neutrons lose their energy of motion to the moderating nuclei, as they bounce off them. The energy transfer is more effective the lighter these nuclei. The most commonly used moderators are graphite, a form of carbon (nuclear mass approximately twelve times that of the neutron), and ordinary or 'light' water, which contains hydrogen (mass approximately equal to that of the neutron). Light water is the most effective at slowing neutrons, but captures most.

The consequences of neutron capture are different for

Below: the blue glow is Cerenkov radiation, caused by the entry of electrons into water at speeds greater than the speed of light in water. *See also pages 108–109.*

different types of nucleus. The nuclei of naturally occurring uranium atoms are of two types of isotopes, containing different numbers of particles. 99.3% have 238 particles, but 0.7% have 235. It is the existence of uranium-235 that has made possible the nuclear reactor.

When a neutron is captured by uranium-235, the resulting compound nucleus, uranium-236, may remain intact, but is more likely to undergo fission, splitting into two fission product nuclei of approximately equal mass, and emitting fast-moving neutrons. Because it is so likely to undergo fission, uranium-235 is called a fissile material.

Capture of a neutron by uranium-238 yields, as a compound nucleus, uranium-239. This does not split, but it undergoes spontaneous radioactive transformations, increasing the positive electric charge of its nucleus by emitting two beta-rays and so becoming plutonium-239. This nucleus, an isotope of the second element beyond uranium in the periodic table is fissile. Because it can be converted into a fissile material, uranium-238 is said to be a fertile material. In nuclear reactors, both fissile and fertile materials have important parts to play.

A chain reaction of nuclear fission proceeds steadily if, on average, one of the two or three neutrons emitted in each fission triggers off a further fission. This leaves one or two spare neutrons per fission. In a nuclear reactor, some of these spare neutrons are deliberately absorbed in 'control rods', partially inserted into the moderator. Others are unavoidably lost by capture in uranium-235 not followed by fission, by capture in the moderator, coolant, fission fragments and structural materials and by leakage from the reactor. The rest may profitably be absorbed in the fertile material, so producing new fissile material.

Some thermal reactors are fuelled with uranium of natural isotopic composition, but, for the majority, the uranium is first put through an isotope separation process, to remove some of the uranium-238. The use of this 'enriched' uranium, usually with between 2 and 4% uranium-235 content, gives greater freedom in reactor design.

In practice, the ratio of the number of plutonium-239 nuclei produced to the number of uranium-235 nuclei consumed in a thermal reactor, is less than one. If the plutonium is required for nuclear weapons, the fuel may be

Bottom of page, left to right: a handling flask for irradiated fuel at Winfrith, England; the device for inserting fuel at Dounreay, Scotland; checking the fuel temperature at Windscale.
Right: the top of the core of a High Flux Isotope Reactor.
Far right: the Advanced Gas-cooled Reactor at Windscale.

discharged at an early stage. Otherwise, it is left in the reactor for a period of between three and five years, during which some of the plutonium-239 is burnt up by fission and some is converted to higher isotopes, plutonium-240, -241 and -242. Of these, only plutonium-241 is fissile.

Thorium

Despite the conversion of some uranium-238 into plutonium, thermal reactors cannot, on their own, afford a prospect of being able to burn up by fission more than about 2% of a stock of natural uranium. It is with the objective of improving the utilization of potential nuclear fuels that two further lines of nuclear reactor development are being pursued. One scheme is to substitute thorium, which has only one naturally occurring isotope, thorium-232, for uranium-238 as fertile material This would result in the production of the fissile material uranium-233. The properties of this nucleus are better suited to thermal reactor operation than are those of plutonium-239 and it is believed that its use would make it possible to burn up a high percentage of a stock of thorium.

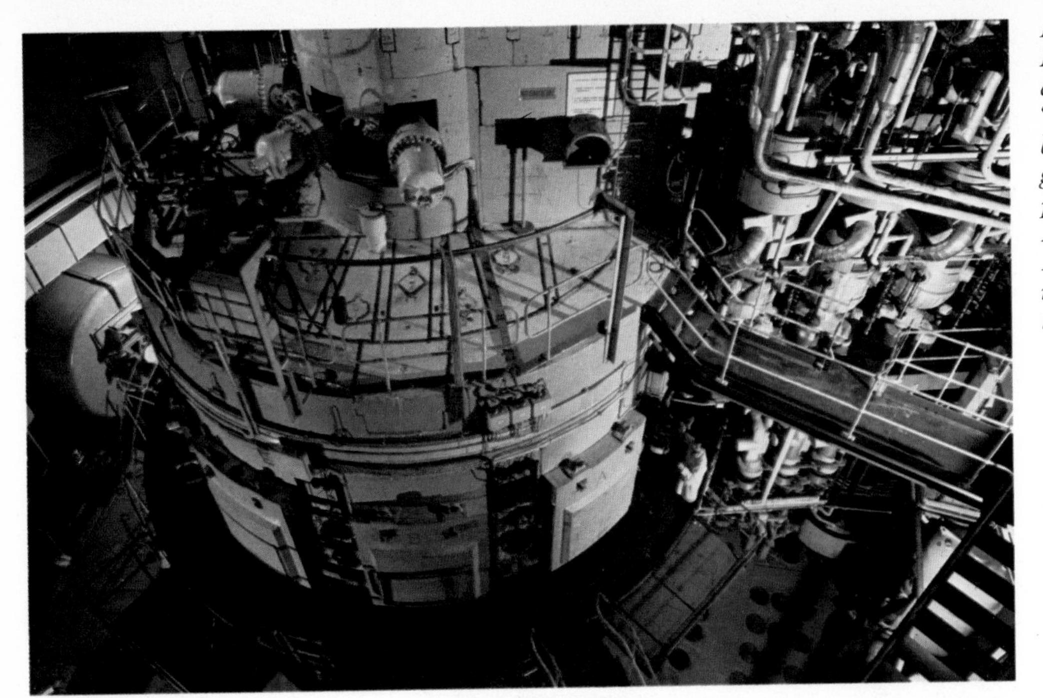

Left: another view of the High Temperature Reactor at Winfrith, England. The 'Dragon' reactor is cooled by helium and modified by graphite; it is one of the projects set up by the European Nuclear Energy Agency, a project started in 1959 and supported by twelve countries.

Fast breeder reactor

The other line of development is the fast breeder reactor, which uses plutonium isotopes as fissile material and uranium-238 as fertile material, with no moderator. Operating with fast neutrons, the properties of plutonium are better, there being less tendency for it to be converted into non-fissile isotopes. The result is that more than one nucleus of uranium can be converted into plutonium-239 for each plutonium nucleus destroyed, so that the initial stock of plutonium grows or 'breeds'. The important point is that fast breeder reactors should afford the means of obtaining very much more energy from the world's uranium reserves than would be possible with thermal reactors alone.

An atomic nucleus is positively charged. When it is undergoing fission, the two parts separate until the forces of attraction associated with binding energy cease to be effective. The two parts, each of which carries about half of the charge, are then driven apart at great speed by electrostatic repulsion. The major part of the energy released in fission initially takes the form of this energy of motion of the fission products. As these are brought to rest in the surrounding fuel, the energy becomes distributed, appearing as heat. There is a further liberation of heat as the fission products, which form atoms of about 30 different elements, undergo radioactive transformations.

Fuel elements

In some reactors, the fuel is uranium metal, but uranium oxide and carbide stand up better to high temperatures and to the accumulation of fission products. Rods or pellets of fuel are sealed into thin-walled metal tubes which, in some designs, are grouped into clusters of 36 to form a fuel element. For the highest temperatures, graphite is used instead of metal for cladding the fuel.

The fuel elements, in general, are held in vertical channels through which flow streams of coolant. The most commonly used coolants are the gases carbon dioxide or helium, water, either 'light' or heavy (containing atoms of the hydrogen isotope deuterium rather than normal hydrogen), and molten sodium metal. When water is used, it is either held under such high pressure that it remains in liquid form at a temperature far above the normal boiling point, or else the pressure is adjusted so that steam generation occurs as the water is passing over the fuel elements. In the latter case, the steam passes direct from reactor to turbine. Otherwise, the coolant goes to a heat exchanger, where it gives up heat to a secondary stream of water, which turns to steam.

Whichever coolant is used, it must be kept under pressure. This is usually done by enclosing the fuel, moderator and coolant in a pressure vessel, but an alternative is to place fuel and coolant only in an array of pressure tubes with the moderator outside. In all reactors, a concrete biological shield several feet thick protects people in the vicinity from radiation.

Nuclear waste

Eventually, mainly because of the deleterious effects of the accumulating fission products, the fuel must be discharged from the reactor. It is next allowed to 'cool' for some months, to let much of the radioactivity die away, and is then transported in heavily shielded 'flasks' to a reprocessing plant, such as the Windscale Works of British Nuclear Fuels Limited. Here, the residual uranium and plutonium are extracted by chemical separation.

Of the fission products, krypton and xenon isotopes are released to atmosphere during reprocessing. Small quantities of other fission products, dispersed in large quantities of water, may be discharged into the sea or, if the reprocessing plant is inland, pumped into the ground. Apart from these, the fission products are stored in jacketed stainless steel tanks, placed in thick-walled concrete 'cells'. Among proposals for the ultimate disposal of these wastes is their incorporation into highly insoluble glassy substances and their entombment in carefully chosen geological formations.

If nuclear power plants continue to be built, especially using breeder reactors, the storage of these poisonous wastes for hundreds of years is going to be a problem. The safe transport of fuel elements between reactors and fuel processing plant is another problem. Fusion, the other means of producing nuclear energy, would be much safer, using deuterium (from sea water) and tritium (from lithium, which is plentiful). So far, however, no practical methods of controlling fusion reactions have been developed.

VEGETATION

The world's forests are being used up faster than they are being replaced, and the fossil fuels formed from prehistoric forests are also disappearing.

Below: felling timber in a Karri forest in Western Australia. Bottom of page: these logs will be left on a frozen river until spring.

The world's annual forest woody increment is estimated to e $12{,}900 \times 10^6$ tons of which only 13% is harvested. The emaining 87% is capable of producing 5×10^6 MW yr, a uantity approaching the present annual world power onsumption. Unfortunately, this source, like the wind, uffers from lack of concentration. The ecological and limatic effects of indiscriminate cutting would be severe ut, given scientific management on a continuous basis, orests could be a useful source of power for some underleveloped countries.

The production of photosynthesized material as an nergy source on a minor scale might be possible using treams and ponds loaded with organic effluent. These can roduce material at a rate comparable with that of good rable land, about 10 times the rate of production in forests. Certainly arable land itself cannot be used for power proluction: it will all be needed for agriculture. The world opulation is expected to increase to about 6×10^9 by the ear 2000, and the energy required for food production will hen be about 1×10^6 MW. If the energy problem is not olved, it could mean a drop in material standards in the leveloped countries, or famine in the underdeveloped. This survey has shown that water power, geothermal ources, and terrestrial solar radiation (whether via solar ollectors or photosynthesis) can make only a minor contribution. The two safe long-lived sources of power, namely usion and space solar energy, are not yet within our grasp.

ELECTROMAGNETICS

MAGNETISM

Magnetism is closely linked with electrostatic force: the difference is that while a stationary charged particle has just an electric field associated with it, a moving particle has a magnetic field as well. Yet permanent magnets—the most familiar type—have no apparent associated electric field. Modern physics has explained permanent magnets in terms of electromagnetism, though as with all scientific explanations, the theories only show the way in which things work, and fail to clarify just why they should do so in the first place.

Faraday, in the 1830s, demonstrated clearly the relationship between electric charge and magnetism. When an electric charge moves it is said to constitute an electric current. When an electric current flows, it generates a magnetic field in the space around it just as if the current system had been replaced by a magnet system with a particular shape. It takes a force, analogous to the pressure needed to cause water to flow in a pipe, to make a charge move, that is, to produce an electric current. This force is known as an electromotive force (emf). Faraday showed that when an object capable of conducting electric current was moved through a magnetic field, an emf was set up in the conductor, capable of producing electric current. He also demonstrated that when the magnetic field which 'threaded' a conducting object was changed, an emf was also produced. So electricity produces magnetism and magnetism produces electricity.

Magnetic circuits

The concept of magnetism existing only in closed loops is a useful one and was known to Faraday. It can be shown experimentally that the 'driving force' in a magnetic circuit of an electromagnet (analogous to emf in an electric circuit) is proportional both to the number of turns which form the coil and the current in that coil. This is called the magnetomotive force (mmf) and measured in ampere-turns. We then invent an imaginary 'substance' which we consider to be the result of this mmf; this we call magnetic flux (or sometimes 'induction') and is measured in webers. Then we can write an equivalent of Ohm's law for a magnetic circuit and use it to calculate the mmf needed to set up a certain flux or the reverse. In electrical circuits emf equals current times resistance, and in magnetic circuits mmf equals flux times reluctance (the impediment to the flow of flux).

Permanent magnets

This tidy pattern of what is universally known as electromagnetism is, however, upset by an aspect of magnetism that has no precise counterpart in electricity. Some elements, having been placed in a magnetic field and then removed from it, adopt and retain an apparent internal source of mmf and they continue to drive a flux pattern in the space around them. These we call permanent magnets and the phenomenon 'ferromagnetism', because one of the elements is iron (Latin *ferrum*). The others are the less common metals cobalt and nickel and still rarer elements such as gadolinium and dysprosium.

Many textbooks on magnetism and electricity begin with the discovery of lodestones in the Chinese desert in 3000 BC. This, and the fact that horseshoe and bar magnets have been available in toyshops for many generations, has resulted in a demand for an 'explanation' of their behaviour at a very early part of a school curriculum. The fact that the Earth itself is magnetized along an axis roughly corresponding to its axis of rotation was used by the ancient Chinese, who found that a freely suspended lodestone would always set itself in the same geographical direction. Simple rules were needed to deal with a simple action-at-distance experience, and soon the rule emerged that the lodestone had 'poles' that pointed north and south. This was followed by the rule that 'like poles repel, unlike poles attract'.

From an educational point of view it is doubtful whether the readily available 'permanent' magnets are a blessing or a curse, for they focus attention away from the circuita

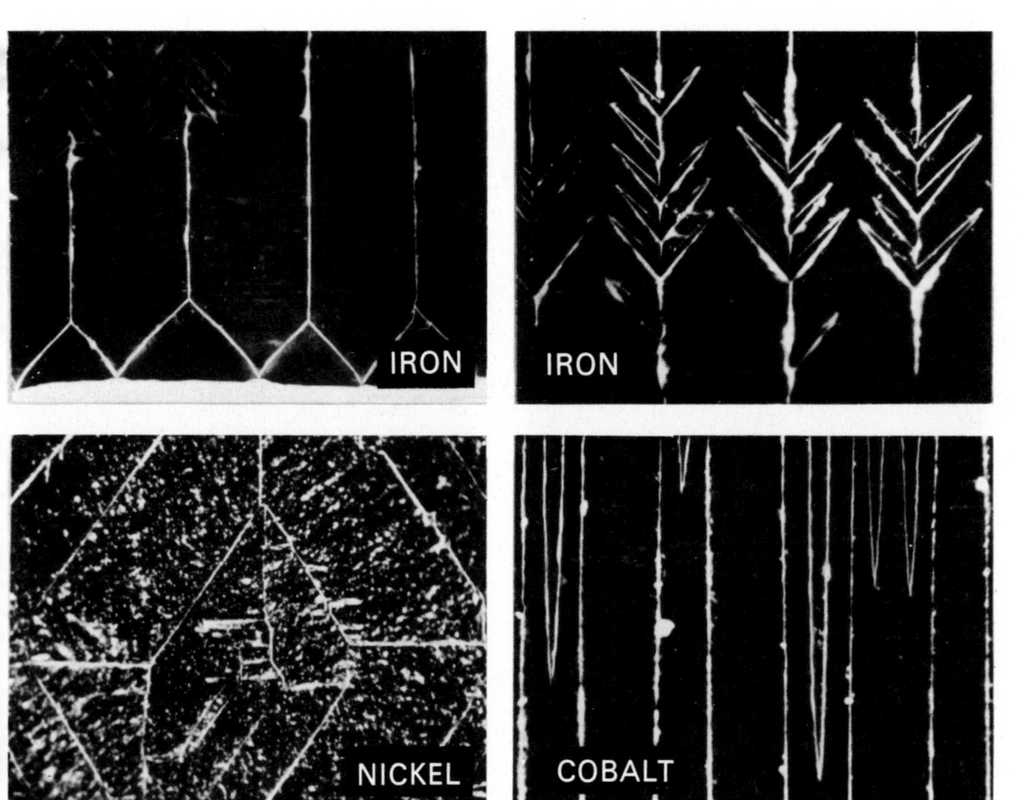

Opposite page: at a few degrees above absolute zero, the pan, made of lead which is normally a poor conductor, becomes superconductive. As a magnet is lowered into the pan, induced currents are strong enough to support it. On this page: magnetostrictive energy in domain structure of pure iron, nickel and cobalt crystals. When magnetized, the crystals change length and create internal strain.

nature of magnetism, which is far more useful in engineering than is the pole concept. Arguments against the circuital concept are that unlike electric circuits, magnetic circuits cannot be insulated because the corresponding magnetic conductivity of air or empty space is finite and only about a thousand times smaller than that of the best magnetically conducting steel. Faraday himself likened the design of magnetic circuits to the design of electric circuits using bare copper wire in a bath of salt water (which is a good conductor of electricity). The result is that while the electric circuits of most electrodynamic machines are complex and consist of coils of thin wire, multi-turned, their magnetic circuits are simple, short, fat and consist of a single turn.

The disadvantages of the pole concept however are first that it presupposes isolated poles in space; these have not been found, despite searches (isolated electric charges, however, do exist); and second, that troubles arise when trying to predict reactions between permanent magnets and other, initially unmagnetized, pieces of ferromagnetic material. For this purpose the law of induced magnetism is invented, implying that the proximity of a primary magnet pole to a piece of soft iron induces an opposite pole in the latter nearest to the magnet and a similar pole at the point most remote from the magnet. There are a number of experiments which are very hard to explain on the basis of poles alone, while the circuit concept sees no difficulty in such arrangements: the pieces will always take up a position of minimum reluctance (that is, minimum impediment to the flow of flux), within the prevailing mechanical constraints. The circuit theory sees a magnet's pole merely as a change in reluctance between different parts of a magnetic circuit.

The hysteresis loop

The difficulty with permanent magnets does not end here. The magnetic flux density B (the number of flux 'lines' crossing unit area—measured as webers per square metre) that can be induced in a ferromagnetic material by imposing an mmf on it is not directly proportional to that mmf. Furthermore, if flux density is plotted as a graph against magnetizing force H (a quantity obtained by dividing mmf by magnetic length and measured in ampere-turns per metre) the state of magnetization of a piece of ferromagnetic material is seen to depend on its entire magnetic 'history'. If taken through many circles until a repetitive pattern occurs, the loop so obtained is called a hysteresis loop whose area can be shown to be the work done in taking the magnetic material around one cycle and which appears as heat in the material itself.

The ratio B/H is given the name permeability (μ) but it is clear that μ is not constant as B varies. It is this aspect which becomes such an easy concept in magnetic circuits involving no ferromagnetic material, for μ is then constant and is the magnetic conductivity directly usable in the magnetic circuit version of Ohm's Law.

Theories of ferromagnetism

The behaviour of ferromagnetics is one of the hardest phenomena to explain in terms of conventional physics. For several generations, Weber's theory that all ferromagnetic substances were effectively made of a mass of tiny particles,

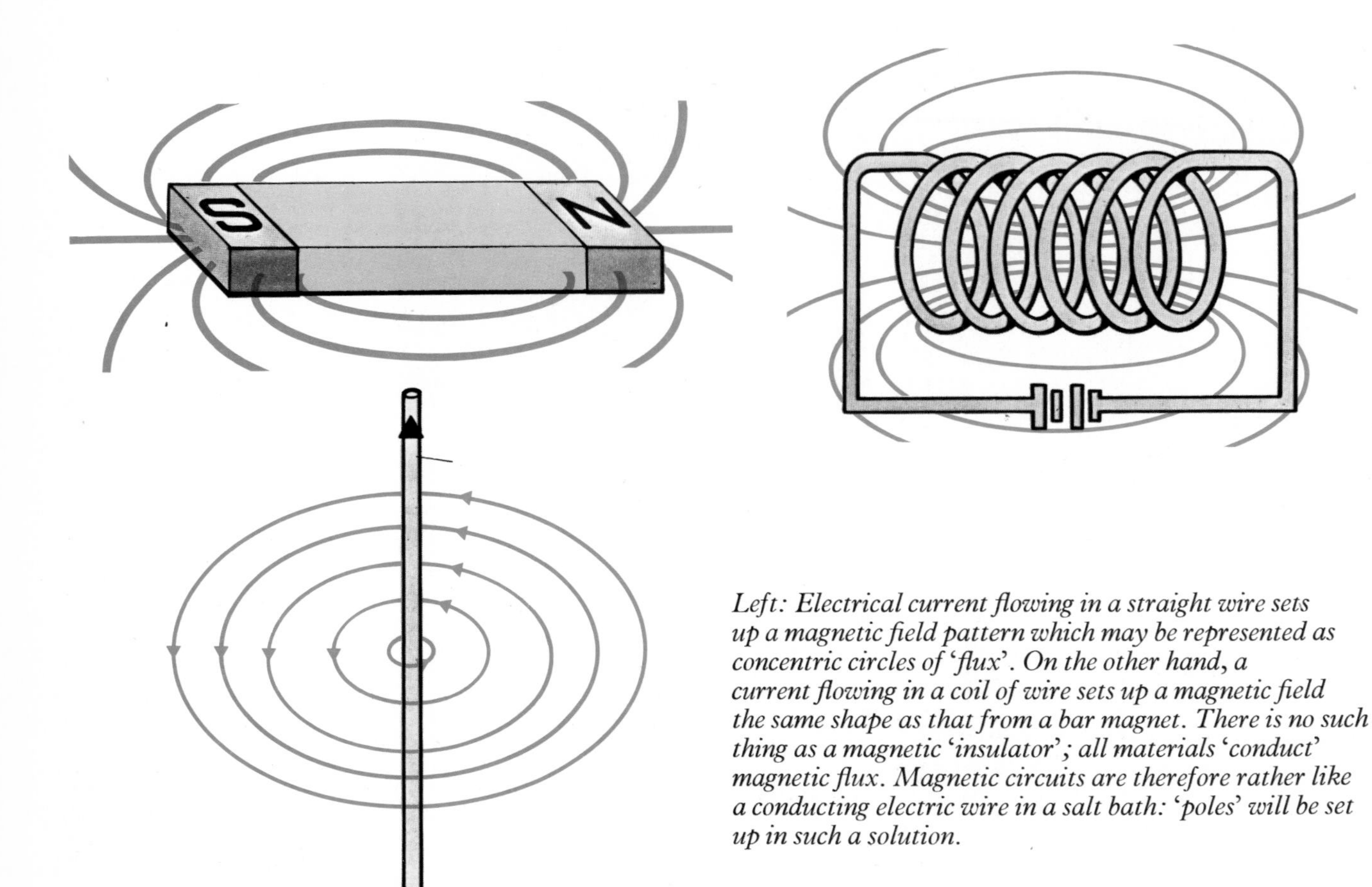

Left: Electrical current flowing in a straight wire sets up a magnetic field pattern which may be represented as concentric circles of 'flux'. On the other hand, a current flowing in a coil of wire sets up a magnetic field the same shape as that from a bar magnet. There is no such thing as a magnetic 'insulator'; all materials 'conduct' magnetic flux. Magnetic circuits are therefore rather like a conducting electric wire in a salt bath: 'poles' will be set up in such a solution.

each of which was a permanent magnet, remained the only plausible model, but the imagination had to be strained to account for various degrees of difficulty being encountered in forcing the 'micromagnets' to line up, as exhibited in the hysteresis effect.

The modern domain theory of ferromagnetism could be said to begin with the observation of the Barkhausen effect, whereby audible 'clicks' were heard in a telephone receiver when connected to a coil of wire surrounding a ferromagnetic specimen that was being magnetized by a very gradually increasing mmf. Most of the pioneering work was done at the Bell Telephone Laboratories in the USA. The technique evolved was to polish the surface of a specimen and then etch it with acid. This revealed patterns of 'walls', later to be called Bloch walls, dividing regions of different magnetic orientation. The theory of these domains was then built up to the following detailed picture.

A clue to the essential difference between a ferromagnetic element and a non-ferromagnetic one is to be found in their atomic structures. The electrons orbiting the nucleus of an atom are arranged in shells, beginning with the simplest atoms 'filling' the innermost shell and shells of large radius then being used for the electrons of elements higher in the periodic table, and therefore with more electrons. In order to balance the forces and energies within an atom it is necessary to credit orbiting electrons with a spin (even though it is sometimes necessary to describe an electron as energy and therefore without shape).

One feature that is unique to the ferromagnetic elements is that their atoms each have at least one electron with an uncompensated spin in one of the outer shells. Latching on to this as the basis of the mechanism, it was soon shown that atoms finding themselves in relative positions such that the axes of the uncompensated spins were parallel, were likely to stay thus related and thereafter to regiment neighbouring atoms in this direction. This build-up process is rapid and implies that every crystal of a ferromagnetic metal structure is self-magnetizing to saturation level.

In the case of each single crystal, however, what had hitherto been defined as the degree of magnetization was not due to each part of the crystal being magnetized to the same fraction of its saturation level. Rather the crystal is divided into portions or domains, each of which is self-saturated, but not all domains are lying in the same magnetic direction. The rules for the domains can be investigated both experimentally by the etching process and theoretically, for the crystal as a whole tries to assume a pattern of minimum total energy.

The various forms of energy in a magnetic metal crystal may be classified as follows. Firstly, there is magnetostatic energy, where the flux has to emerge from the crystal and pass through space. Secondly, there is magnetostriction energy due to the crystal increasing or decreasing in length because of magnetization, producing mechanical strain energy. Thirdly, there is Bloch wall energy due to strain in

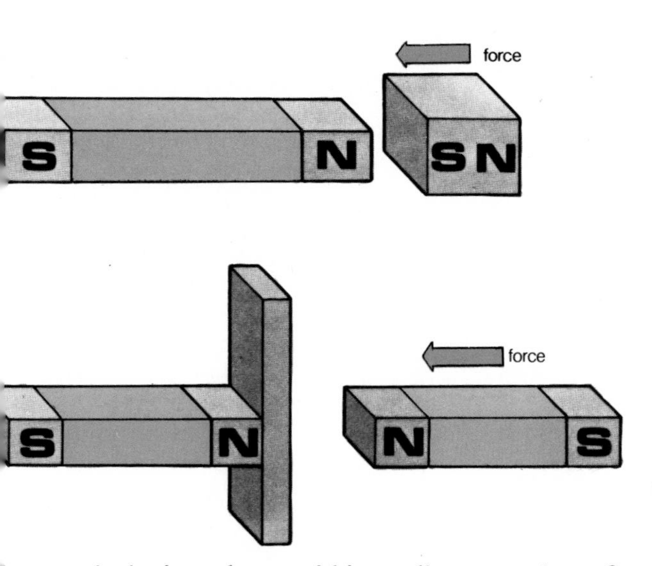

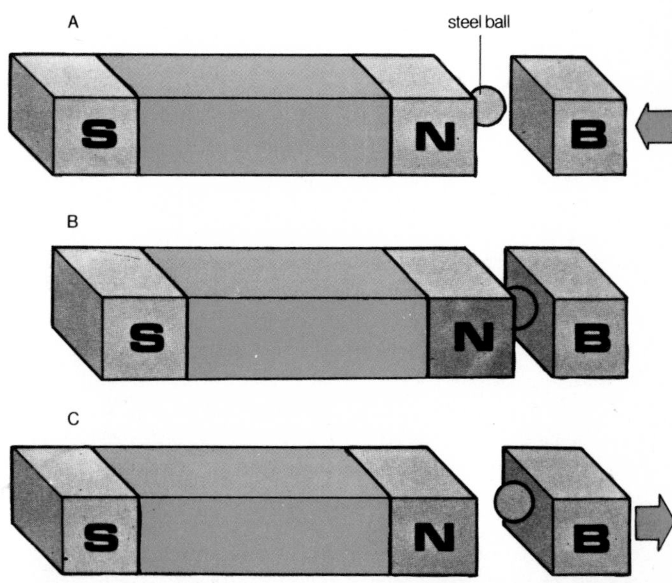

he atomic lattice where, within a distance only a few toms thick, the spin must reverse.

When a ferromagnetic metal is reduced to finer and ner particles, the minimum energy condition is reduced to nagnetization in one direction only. This knowledge gave ise to a whole new magnet technology in which fine netallic powders were shaken so as to allow them to line p with each other (just as Weber's theory stated) and the vhole then heated (sintered) to make a permanent magnet vith a very large coercivity. (This is the magnetizing force equired to 'pull' the flux down to zero after its initial nagnetization). This technique can be extended to include ubstances other than pure metals, known as ferrites. These re ceramic materials consisting basically of iron oxide and mall quantities of transition metal oxides such as cobalt nd nickel. It was found that in the case of non-metallic nagnetic materials not all the electron spins within a omain were aligned in the same direction. Instead, a ortion of the atoms had their uncompensated spins ligned in one direction while the rest of the atoms had neir spins aligned in the opposite direction. Externally, nerefore, such materials cannot contain as high a flux evel as can metallic substances. These ferrite substances re said to display ferrimagnetism. They are used in micro-vave apparatus where it is essential that the substance is not apable of conducting electricity. Furthermore, the flux aturation level can be controlled by mixing different naterials.

More recent developments have produced flexible and ven liquid magnetic materials by the powder-metallurgy rocess, suspensions of barium ferrite and other similar naterials being held in a base of rubber, polyvinyl chloride nd other plastic materials. In the case of liquids, each article of ferrite is encapsulated in a single layer of mole-ules of a long-chain polymer. The capsules slide over each ther with virtually no frictional resistance, so that when nade as a suspension in water, the liquid has the viscosity f water but the particles are so small that thermal agitation revents them from settling.

Above top: some magnetic phenomena cannot be explained by the induced pole concept, but by circuital theory: the system will always tend towards a position of minimum reluctance. A permanent magnet is said to induce poles in an initially unmagnetized piece of soft iron; the end nearest the north pole becomes a south pole and the iron is attracted. But when a thin strip of unmagnetized steel is attached to the north pole of one magnet, there is an attractive force on the north pole of another magnet. Above right: the steel ball, initially unmagnetized, can always be pulled off the magnet's pole face by touching it with an unmagnetized piece of iron (B). The 'circuit' has a minimum reluctance when the ball is attached to (B). Above: permanent magnet in a sample of lodestone.

ELECTROMAGNETISM

If a permanent magnet attracts a piece of iron or steel, that is a purely magnetic action. If a battery sends electric current through a wire so as to heat it, that is an electric effect. But wherever an action takes place involving both magnetism and electricity, such action is said to be electromagnetic. There are many manifestations of this phenomenon, which was first discovered by the Danish scientist Oersted and greatly enlarged by the subsequent work of Faraday in the first part of the nineteenth century.

One common manifestation of electromagnetism is that a current flowing in a wire produces a magnetic field—this is the operating principle of an electromagnet, and can be harnessed to produce motion in electric motors through the attractive and repulsive forces of magnetic fields. When a magnet (either a permanent magnet or electromagnet) is moved near an electrical conductor, turbulent eddy currents are induced in the conductor and it experiences a 'dragging' force. This dragging force can be used to produce motion, and conversely, the eddy currents can be harnessed to produce a useful electric current (such as in alternators and dynamos). This is an example of a moving magnetic field producing an electric current.

A more complex example of electromagnetism is found in devices such as transformers where a changing magnetic field produces a current. Here, two coils of wire are placed close together. When a changing current (changing in amplitude and/or direction) flows through one coil a changing magnetic field is produced, which induces a voltage in the second coil. If this second coil is included in any kind of electric circuit a current flows.

Understanding by analogy

These phenomena are not fully understood by man. But in order to exploit them, we devise mental models called 'analogues' to help us to obtain at least an appreciation and a hope that through this means we may learn to design better machines by using a phenomenon which is no more understood than is gravitation.

For electric circuits we imagine that electrons flow in wires in much the same way that water flows in a pipe. We know that pressure is needed to make water flow so we invent an electrical pressure and call it electromotive force (emf) or voltage. The convenience of this analogue is that it allows us to use the equivalent of the frictional resistance in the water pipe which increases in proportion to the length of the pipe but decreases in proportion to its cross sectional area. Then, by another analogy, we can invent a magnetic circuit, in which the driving pressure is called magnetomotive force (mmf) and the 'substance' which it

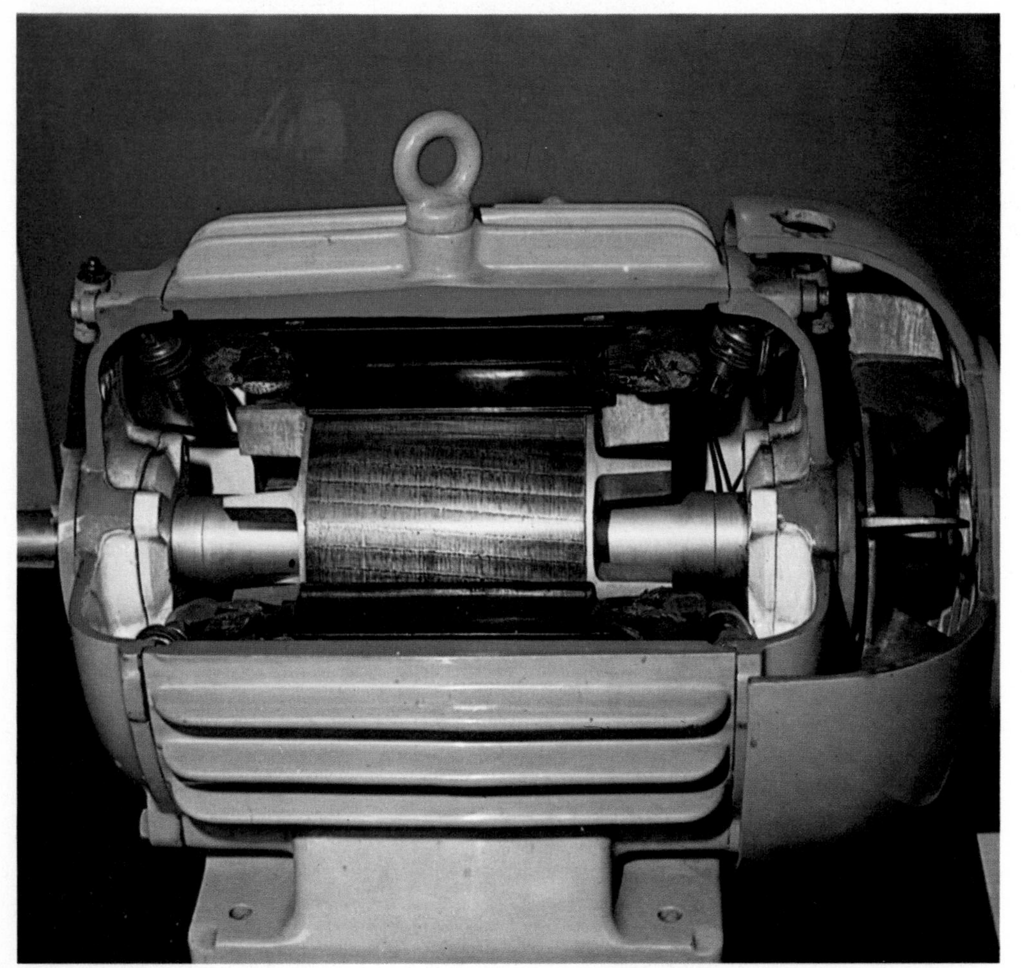

Left: a cutaway view of a three-phase squirrel cage rotor induction motor. Turning of the shaft is 'induced' by an electric current. The field in a motor will rotate once with each cycle in the current; thus, in Europe, where the mains supply is 50 Hz (Hertz, or cycles per second) an electric motor cannot run at more than 3000 rpm. (In North America, where the supply is 60 Hz, the figure rises to 3600 rpm.) Another draw-back of an induction motor is that it cannot provide efficient speed variation without a variable supply frequency. It also 'slips', that is, the speed is never quite as high as the frequency of the mains, and it varies with the load on the shaft. A synchronous motor will give constant speed, but is not self-starting; a commutator motor will give variation but has brushes on the segments and needs more maintainance.

ELECTRICITY

rives around the circuit is even less 'real' than the flow of lectrons in an electric current. We call it magnetic flux. Many authors and teachers declare that, despite its name, ux does not flow. The fact is that it does not exist, except s a human concept, and the only 'right or wrong' about its ow is to be judged on whether the concept is useful to a articular individual. For some, it is more profitable to hink of flux as merely being 'set up' because it represents nly stored energy, and not a continuous loss of power as is he case when electric current flows in a wire. For others, he analogue is more profitable if flux is considered to be a nore precise analogue of electric current so that a magnetic ircuit can be given the properties appropriate to an electric ircuit.

Linking electric and magnetic circuits

When discussing electric motors, generators and transormers, it is essential to note that each machine includes t least one electric and one magnetic circuit. Since there is no simple equivalent in magnetic circuits to the insulating naterials of electric circuits, it is usual to design a machine vith only one magnetic circuit but two or more electric ircuits. Moreover, for the same reason, electric circuits in nachines are usually multi-turn coils of relatively thin, nsulated wire. Magnetic circuits tend to be single-turn, hort and fat.

The subject of electromagnetism can therefore be xpressed as the linking of electric and magnetic circuits. n such a linking, the driving pressure from one circuit is een to be derived from the flow in the other, and viceersa. For example, in a transformer an alternating voltage emf) across the primary windings produces an alternating urrent in the windings. This produces an alternating mmf n the magnetic circuit, which creates an alternating flux. The alternating flux induces a voltage in the secondary vindings, which, if connected in an electrical circuit, roduces current.

Vector quantities

The commodity we seek to produce in an electric motor is orce which arises as the result of multiplication of flux by urrent, but it is no ordinary multiplication, for the only quantities of flux and current which are effective are those which cross each other at right angles. Quantities which have both magnitude and direction are called vectors and when determining the interactions of vectors with each other the direction as well as the magnitude must be taken nto account. In the above example, the force vector is the esult of the vector multiplication of the flux and current ectors. Where the flux and current vectors are not at right ngles to each other they must be resolved into parallel and ight angular components, but it is always the right angular omponents which produce the force vector. Furthermore, he force vector is always at right angles to both the flux and urrent vectors.

Vector multiplication and, more generally, vector mathenatics is only a form of 'shorthand' for handling quantities which have been shown experimentally to interact in this nusual way. This is another example of an analogue, useful because it works, even though we are not sure why.

Above: lack of experience made early experiments with electricity both exciting and dangerous. This engraving shows George Richmann after being struck by lightning.

Electricity, as its name suggests, is intimately bound up with the properties and consequent behaviour of the electron. The charge on the electron and its relatively high mobility, the ease with which it can be caused to move, are its most important properties from the point of view of electricity. It is the motion of electrons that lies at the root of all electrical phenomena, as, for example, in a current in an electric circuit.

Electrons are found in all atoms, surrounding the compact core, or nucleus, of protons and neutrons in orbital layers or 'shells'. The electrostatic attraction between the protons and electrons holds the atom together. The electrons in the outermost shell, which are often termed the valency electrons (those which take part in chemical bond formation), are only weakly bound to the nucleus in many atoms. This is especially the case with the valency electrons of the atoms of the metallic elements, such as copper, silver, and sodium. The weakness of the attraction, combined with the fact that the mass of both the proton and the neutron is each nearly two thousand times that of the electron, means that the valency electrons can be relatively easily moved. Since electrons are present in all atoms, electricity, broadly

Below: a sheet of Perspex (Plexiglass) was charged by scanning a high energy beam of electrons across it, then earthed (grounded). This spectacular discharge pattern occurred.

defined, is found in many forms throughout nature. Photoelectricity, the movement of electrons from their atomic levels by the action of light, and the whole field of electrochemistry are examples of the extent to which the effects of the electron's properties are encountered.

Current

When the atoms or molecules of an element or compound come together, they arrange themselves in a regular array called the crystal lattice. The type of crystal lattice and the mobility that it allows the valency electrons, and the nature of the particular atom itself, largely determines the electical properties of the element or compound. Resistance, for example, is a measure of the amount of energy expended by the electrons in moving against the opposition of the atoms in the lattice. This expended energy appears mostly in the familiar form of heat. Capacitors exploit the low electron mobility of insulators to build up a surplus of electrons on one of their plates and a deficiency on the other. This imbalance of charge can later be 'discharged' in a short pulse of current.

The lattices of metallic elements allow the valency electrons a high degree of freedom and present little opposition to their motion. As a result, they are excellent conductors of electrical current. Insulators, on the other hand,

Below: organic radioactive compounds are stored in liquid nitrogen to minimize the effect of the radiation on chemical decomposition of the compound.

are characterized by the low mobility their atoms and lattices allow the valency electrons.

The crucial importance of the type of crystal lattice is demonstrated by the case of the two distinct crystalline forms, or allotropes, of pure carbon. In the graphite form, carbon is an excellent conductor, but in the diamond form it is a poor conductor. The atoms making up the two lattices are exactly the same, but the electron mobilities allowed by the lattice types are completely different.

Individual electrons move along solid conductors in a random 'shunting' manner; it takes a flow of over six million million million electrons per second to make a current of one ampere, named after the French physicist A M Ampère. While the effects of a current are transmitted with the speed of light, each electron is only drifting in the direction of the current at an effective rate of around one inch per second (2.54 cm per sec). When ionic compounds, such as common salt are dissolved in water, the water molecules cause the sodium and the chlorine atoms to split up, or 'dissociate', into free ions. The negative chloride ion, which carries the single valency electron in the outer shell of the sodium atom, and the correspondingly positive sodium ion are then able to carry a current through the solution. Instead of the mobility of the electron, it is the mobility of the ions through the solvent (in this case water) that determines the resistance of the solution.

Voltage

Although the electrostatic attraction between the proton and the electron is one of the fundamental forces of the Universe, the systematic motion of electrons that constitutes a current is not directly caused by it. Current flows between points of 'potential difference', this difference being expressed in volts. The voltage between two points is a measure of the energy an electron gains in moving them: the electrons move because it is preferable for them to do so, and the energy arises as a result of the 'energy levels' the electron can fill in the atom.

The complete theory of electricity, which clearly hinges around the atomic and quantum theories of matter, has only evolved during this century. But the scientific observation of electrical phenomena, and even their exploitation in practical applications such as the battery, long preceded this full understanding. This is another example of a common occurrence in science: the laws of nature have often been put to good use even though the principles underlying them were either unknown or misunderstood.

Triboelectrification

The first record of the study of electricity can be found as early as the sixth century BC. A Greek, Thales of Miletus (624—548 BC), observed that when certain substances were rubbed they were able, for a time, to attract light objects. Most people are familiar with this phenomenon in one form or another: the effect a dry nylon comb has on hair, making it stand on end, is one example. Also when the plastic body of a fountain pen is rubbed with a silk, cotton or woollen cloth, it can pick up small pieces of paper. This electrostatic effect is also called triboelectrification, and it is one of the most fundamental results of the electrical properties of matter.

The triboelectrification of an object is due to the frictional force applied in the rubbing process. Certain types of atoms can be ionized like this, being either partially stripped of their valency electrons or given some of the valency electrons of another atom. The cloth that is used to induce this ionization, as could be expected, is also ionized— with the opposite charge. It, too, has the ability to attract light objects for a while. This ionization, however, is not all extensive, and the normal balance of positive and negative charge is soon restored when the object is 'earthed' (grounded). This can be done by the action of the moisture in the surrounding air or by being simply touched.

Magnetism and polarity

The word 'electron' is actually derived from the Greek word for the natural resin amber, *elektron*. Amber was one of the first substances to be triboelectrfied. The Greeks connected this phenomenon with magnetism, which they had observed between pieces of magnetite ore. Since both this force of electrostatic attraction and the force of magnetism act 'at a distance'—that is, do not require any physical contact to pull objects together—they seemed to fall into one class. Thus, from the outset, the scientific studies of electricity and magnetism were linked, although for a spurious reason. The first genuine connections between the two were not made until 1820, when Oersted chanced to see that a compass needle was deflected by the presence of a current-carrying wire. Strangely enough, one school of Greek thought did develop a simple version of the atomic theory, but the link between electricity and the

atom did not come until the turn of this century.

In 1551 Jerome Cardan studied the similarities and differences between magnetic and electrostatic attraction. He proposed that electricity or, rather, the triboelectric aspect of it that he was concerned with, was a type of fluid. All things considered, it is not such a bad model. What was more important, it meant that this mysterious force was now treated as something quite material and not at all magical. Fluid theories of electricity became popular in the 18th and early 19th centuries.

In 1600 William Gilbert, an early investigator into electrical and magnetic phenomena, conducted extensive experiments into triboelectrification. He classified substances into good or poor electrifiers, depending on how easy it was to induce in them this ability to attract light objects. His classifications can be seen to correspond to the modern ones of 'insulators' and 'conductors' respectively.

The fluid theories of electricity were subsequently developed further by the French scientist du Fay. In 173 he proposed that there was only one fluid responsible. I 1747 Benjamin Franklin, in his celebrated experimen with kites flown during thunderstorms and an early type c capacitor called a Leyden jar, proposed that there wei two types of electrical fluid. One was responsible fc 'positive', the other for 'negative' electrification.

This distinction between positive and negative cam about because some electrified substances attracted eac other, whereas others were repelled by them. This again another obvious parallel with magnetism: a north pole of magnet repels another north pole, just as a south pole repe another south pole, but a north pole and a south po attract each other. The fluid theories, however, were sul sequently discarded with the advances in scientific know ledge in the 19th century, but the 'positive' and 'negativ terminology was retained, along with the 'like repels, unlik attracts' rule.

Left: the 'flashover' discharge created during testing of a high voltage insulator of the type used in power stations and switching compounds.

Coulomb's law

The quantitative development of electrical theory began in the second half of the 18th century. The independent discovery, by Priestley in 1767 and Coulomb in 1785, of the so called Coulomb law, and its exact expression in a mathematical formula, started it. The Coulomb law is to electricity what Newton's Law of gravitation is to physics, and in fact it is remarkably similar in its general form.

Coulomb's law relates the electrostatic force of attraction or repulsion between two groups of charge to the amount of charge, the distance separating them, and the medium in which they are situated. The force was found to vary with the reciprocal of the square of the distance between the charges. This means that doubling the separation will reduce the force to one quarter of its original magnitude. The electrostatic force is thus a 'short-range' force, as it dies away in intensity very rapidly with distance. This formula has stood essentially unchanged to this day, a tribute to the experimental precision of both scientists.

From then on the understanding of electricity broadened rapidly, due mainly to the invention of the first reliable source of continuous current. This was called the 'voltaic pile', and it is the forerunner of the modern battery. In fact, it relied on the same current-generating process as the modern battery—the controlled electrolytic action of a weakly acidic solution on metals. Its invention in 1800 by the Italian physicist Alessandro Volta, after whom the 'volt' is named, was a turning point in the study of electricity. Before the voltaic pile, the only way of 'storing' charge was a bank of electroscopes (devices for storing electrostatic charges), the current from which was small and usually too short-lived to allow its effects to be studied in any detail. The electroscopes were each laboriously 'charged' by repeatedly transferring the charge from a triboelectrified substance such as amber or glass on to the electroscope.

Conduction

An electric current flows when charged particles move under the influence of an electric field. An electric field is a voltage gradient—that is, it can be considered as a slope down which the current flows—and is measured as volts per unit distance, typically volts per metre. There exists for example a voltage gradient (electric field) between the terminals of a battery. Moving the terminals closer together or increasing the voltage would increase the gradient, and hence the field strength. Just as a ball rolls faster down a steeper slope, this makes the charged particles accelerate faster.

Within an electric field, free negatively charged particles are accelerated towards the positive electrode (anode) while positively charged particles move towards the negative electrode (cathode). Where both types of charges can move, the total current is the sum of the two. Electrons are extremely small and mobile particles carrying negative charges and are the predominant particles involved in most conductive processes in materials. Protons, which are particles carrying positive charges, are larger and about 1800 times more massive than electrons. Furthermore, in most atoms protons group with neutrons (having no electrical charge), making them even more massive and difficult to move compared with electrons.

Most materials, whether they are solids, liquids or gases, consist of atoms and molecules in a stable and neutral form, that is, a nucleus of protons and neutrons surrounded by the correct complement of electrons to exactly balance the number of protons. But, if atoms were so stable, it would seem impossible for conduction—which is the free movement of charges in an electric field—to occur.

Electrical conduction in materials occurs in three ways. Either the electrons can be torn away from their parent atoms (ionization—this happens mainly in gases), or, as in ionic bonded chemicals, the atoms already exist as fairly loosely bonded ions which can be broken apart under the influence of an electric field. Finally, in metals, the electrons exist in a 'cloud' and are easily moved by an electric field.

The particle or 'ball like' description of charges is insufficient to explain all the characteristics of electric currents in certain materials and for a satisfactory explanation it is necessary to consider more complex physics, such as quantum mechanics.

Some materials conduct electricity more easily than others, which leads to the two relative terms: conductors and insulators. Except for superconductors (described below) there is no such thing as a perfect conductor, a material through which charges can move with no impediment, or a perfect insulator, in which it is impossible for charges to move at all.

Conduction in gases

When the terminals of a battery are brought close together, a spark is created between them. This is the visible sign of an electric current being conducted through the air. Preceding this, however, is a chain of 'invisible' events which are crucial in the formation of a spark.

Air, for example, consists mainly of neutral gas atoms and molecules. Because of the motion of the gas atoms, however (that is, the thermal energy related to the temperature of the gas), certain collisions are constantly occurring with sufficient energy to knock off electrons from the outer orbits of the atoms. This creates free electrons and positive ions (positive because they have lost electrons). In an electric field, the movement of these positive ions and electrons constitute a small electric current (maybe less than one million millionth—10^{-12}—of an amp).

As the electric field strength is increased, electrons are accelerated to higher speeds which are eventually sufficient to knock off further electrons from atoms with which they collide. This produces a snowball or avalanche effect, producing more electrons and positive ions, and the current increases rapidly.

Furthermore, some of the positive ions reaching the cathode have sufficient energy to knock electrons out of the metal, increasing the discharge current still more. At the same time, the gas becomes more excited, the temperature increases, and some of this energy is emitted as photons of light. The discharge then becomes visible.

If the air pressure between the two electrodes is reduced,

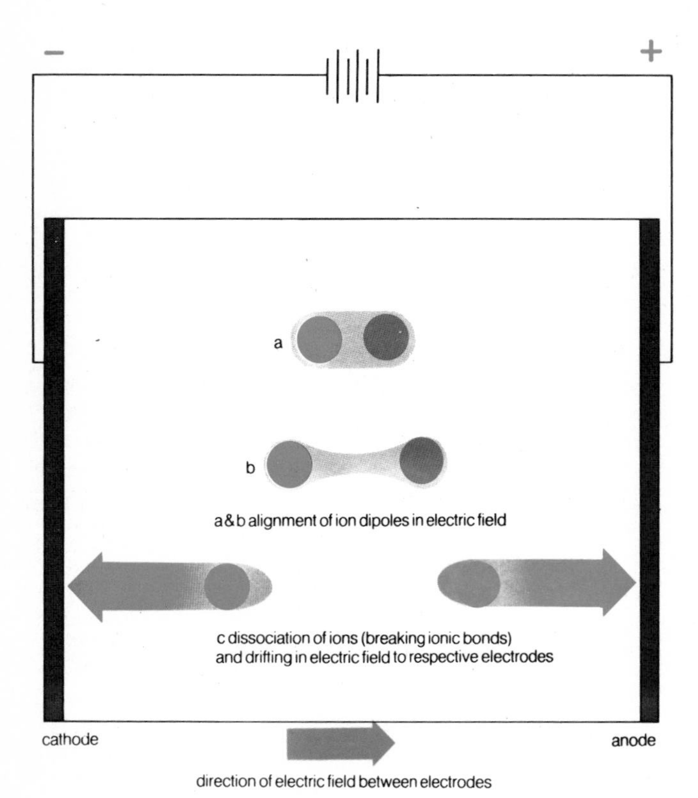

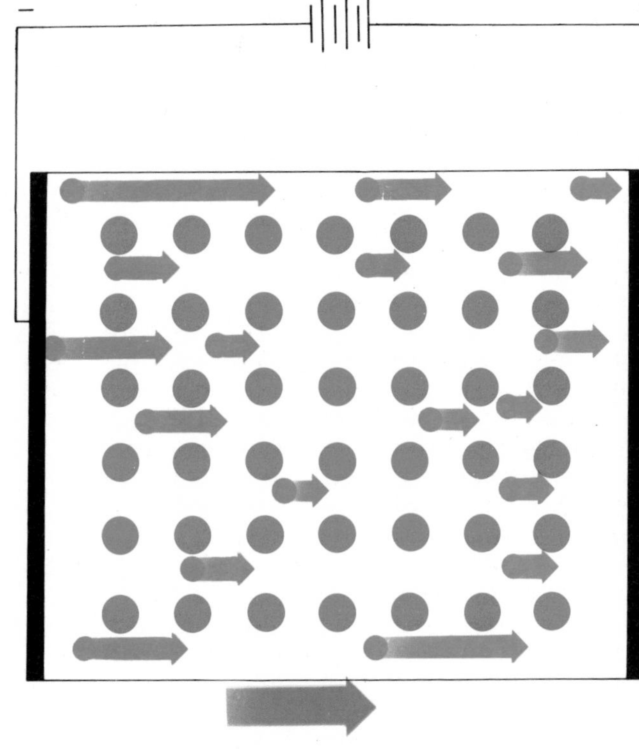

Above: conduction in a metal solid. Metals are good conductors because usually one or two electrons in the outer orbits of the atoms are free to move. These electrons form a 'cloud' which moves almost unhindered.

−
+
+ve ion knocking
electron off cathode
avalanche effect
ionised atom
(one electron missing)
avalanche effect
ionised atom
(one electron missing)
neutral atom
cathode
anode
direction of electric field

Above top: conduction in a solution. When ionic materials are dissolved in water and an electric field applied, the bond can be broken. Above: current builds up in a gas because of the motion of electrons until a spark is seen.

a much lower potential is required for a current to flow. The reason for this is that once an electron has been freed (in some way or other) it has the opportunity to be accelerated to higher speeds by the electric field, because the low air pressure effectively reduces the number of air molecules which impede its progress. Collisions then occur with sufficient energy to ionize the atoms and produce a diffuse glow. This phenomenon is exploited commercially in discharge tubes, used in advertising signs, street lights and so on.

Ionic materials and solutions

In a common salt crystal, negatively charged chlorine ions are bonded to positively charged sodium ions. The application of an electric field across the crystal will tend to move the positive ions towards the cathode and the negative ions towards the anode. The movement is very slow though as the ions are held fairly rigidly together in the crystal structure. If the salt is heated and melted, however, movement is much easier and the current consequently greater.

Another way of increasing ion mobility is by dissolving the crystal in, say, water. Very pure water is a good electrical insulator as its molecules are not easily dissociated into hydrogen and oxygen ions. It acts, however, as a medium which encourages the dissociation and mobility of any ionic material dissolved in it.

GENERATION AND TRANSMISSION

Metals

Metals are excellent electrical conductors. Their atomic nuclei form regular (crystalline) structures surrounded by what is called an electron 'cloud' or 'gas', which permeates the whole structure. This term arises because the electrons which come from the outer orbits of the metal atoms (one or two from each atom) are extremely mobile—rather like the molecules of a gas—and very large currents can flow under the influence of even the smallest electric fields. They also interact with light and endow metals with their characteristic lustre and opacity.

Electrons can move through a perfectly regular crystal without the slightest hindrance. Any irregularities in the structure, however, lead to scattering and a corresponding reduction in conductivity. The most general irregularity is caused by thermal vibration of the atoms (due to heat). As the temperature increases the vibration increases, the electrons are scattered more frequently and the conductivity drops. The presence of impurities or the addition of alloying elements also reduces the regularity of the crystal and decreases conductivity.

A metal which is to serve as an electrical conductor should contain the most mobile electron 'gas' possible, show minimum resistivity due to the thermal vibrations of its atoms and be easily purified. On the first count sodium metal might be a good choice, but its electrons are particularly susceptible to thermal scattering. At room temperature, and above, silver shows the best conductivity with copper a close second. These two metals can also be readily purified and many thousands of tons of copper with a purity of better than 99.97% are produced annually.

Insulators and semiconductors

In an electrical insulator all the electrons are strongly bound to their parent atoms and none is available to form a conducting 'gas'. With semiconductors, some electrons in the atoms are not quite as strongly bound as in insulators. The thermal vibrations of the atoms persuade a few of the electrons to form a mobile 'gas' and hence contribute to conductivity. Compared with metals the conductivity is fairly low but (unlike metals) it increases as the temperature is raised and more electrons are thermally excited into the 'gas'. At a given temperature the addition of specially selected impurities can be used to increase the concentration of conducting electrons and hence the conductivity. Transistors are made from semiconductor materials.

Superconductors

If some metals and alloys are cooled to within a few degrees of the absolute zero of temperature (—273°C) they suddenly lose all electrical resistance and once a current is started in a circuit it will continue to flow indefinitely. This is called superconductivity and has many potential applications. The phenomenon is, however, limited by the technological difficulties of maintaining the extremely low temperatures necessary. The strong magnetic fields associated with the very high currents which can be obtained in superconductors are especially useful and are under development as a possible means of levitation for high speed trains and similar uses.

The first power stations and rudimentary distribution systems were built in the 1880s to supply local lighting loads. Generators were belt-driven by reciprocating steam engines and generation voltages were either DC, or AC of different frequencies. The major advance in power station design was the development of the practical steam turbine by Parsons in 1884. Turbines rapidly superseded steam engines and have undergone continual development. Deptford Power Station, built in 1889 by Sebastian de Ferranti, had four alternators with a total capacity of 1650 hp, 1.23 MW (million watts) and was notable for its high voltage cable to London operating at 10 kV (10,000 volts). As with the turbine, boiler design has advanced from hand firing to automatic fuel supply and boiler control, and now nuclear reactors have replaced conventional boilers as steam generators in many new power stations.

The electricity supply system must be operated to meet consumer demand at any time. Electricity cannot be stored in bulk and station output is constantly adjusted to meet daily variations in load. Power stations are called on to meet the demand in an order of merit determined by operating costs. The cheapest stations operate continuously, apart from maintenance periods, supplying the base load. The base load is the minimum experienced by the system when demand is lowest. Peak load is when demand is highest. As load increases during the day, less economic stations are brought on to the grid (national high voltage distribution system) until at peak the least economic are operating. Two distinct types of power station are required. The large modern power stations with unit sizes of 500 to 1300 MW are economic to run, being very efficient and able to burn cheap fuel. Their high efficiency is obtained through maintaining fine clearances between massive components such as turbine rotors and casings. Load changes must be undertaken slowly to preserve these clearances during differential expansion of the rotors and casings, as the turbine temperature changes when the load is altered. These units are suited for base load generation. Demands above base load are met with increasingly smaller, less efficient, but more flexible steam turbines. Peak load power stations are the opposite of base load ones; cheap to build and very flexible to operate but expensive to run. Gas turbines (similar to jet engines) are typical peak load equipment, their lightweight alloy casings permitting rapid heating and full load within 3 minutes of start. Hydro stations can be either base load or peak load depending on their type and the availability of water. They are most expensive to build but have the lowest operating costs.

Alternators

Alternators are the common feature of all power station plant. They have developed from 60 MW post war to 660 MW (UK) and 1300 MW (USA). Alternating current at either 50 Hz (UK) or 60 Hz (USA) is standard. These frequencies require synchronous speeds of 3000 rpm and 3600 rpm respectively for two pole rotors and half that for four pole machines. In a 660 MW alternator electricity is generated in the stator at 22 kV and 19,000 amps. Currents

Below: the turbine hall of a 2000 MW coal-fired station, which has four 500 MW turbogenerator units.

of this magnitude flowing in the conductors generate heat and it is developments in alternator cooling that have permitted the modern set ratings without overheating and damaged insulation. The alternators are filled with hydrogen under pressure, as it has better heat transfer and lower aero-dynamic losses (power lost due to the resistance of the gas to the motion of the rotor) than air. Water is circulated through the stator windings to cool them directly, and this water is removed through insulated pipes to external coolers. Water cooled rotors will be used on the larger sets.

The alternator is connected to two transformers. The unit transformer transforms some of the power to a lower voltage to drive the auxiliary equipment (such as pumps) associated with the turbine and boiler. The generator transformer transforms the rest of the power to grid voltage for supply to the load centres. For a typical 660 MW generator operating at 22kV, the unit transformer has a secondary voltage of 11 kV and the generator transformer a secondary voltage of 400 kV, the primary voltage in each case being the 22 kV from the generator. In the case of a typical 60 MW set, the generator voltage is 11 kV and the unit and generator transformer secondary voltages are 6.6 kV and 132 kV respectively.

Coal, oil and natural gas

Coal, oil and natural gas fired power stations (often termed 'conventional' stations to distinguish them from nuclear or gas turbine stations) are essentially similar, apart from the detail associated with the handling facilities for the particular fuel. To produce steam the water must be boiled, requiring an input of latent heat. Latent heat is the heat required to change the state of a fluid (for instance from liquid to gas) and it does no useful work. When the steam is condensed in the turbine condenser the latent heat is passed to the cooling water: this heat represents the major loss from the cycle, approximately 35 to 40%.

Superheated steam raised in the boiler is fed to the turbine which is coupled to an alternator. Steam is exhausted from the turbine at low pressure, condensed and pumped back to the boiler under pressure. The temperature and pressure of the superheated steam are made as high as possible, compatible with the safety of available materials, to give maximum cycle efficiency. With austenitic steel (containing 18% manganese, 3% chromium and 0.5% carbon) superheaters on 660 MW boilers generate steam at approximately 2400 psi (165 bar) and 566°C (1051°F). Power station boilers are of massive construction—for example 60 ft (18.3 m) wide, 130 ft (40 m) deep and 200 ft (61 m) high. Water circulates through tubes which completely line the side of the boiler.

The water first enters the economizer section where initial heating to almost boiling occurs. It then passes into a steam drum at the top of the boiler. From there the water recirculates through the evaporator section tubes lining the furnace and back to the drum, where 20% boils off and leaves the drum through steam pipes in the top of the drum. Superheater tubes mounted in the hottest part of the boiler give final heating to the steam.

The turbine has three sections on a common shaft, the high pressure cylinder, the intermediate pressure cylinder and one or more low pressure cylinders. Each section is a purpose built turbine designed to extract the optimum power from the steam at the particular steam conditions prevailing. Steam exhausted from the high pressure section is taken back to the boiler, where it is reheated to its original temperature. In this way more energy is put into the cycle without requiring further latent heat, thus improving the overall efficiency. From the reheater the steam goes to the intermediate section, exhausts directly to the low pressure section, and then passes to the condenser which is operated at near vacuum; this low pressure gives high cycle efficiency. The condensate, condensed steam, is pumped through feedheaters to the deaerator, which removes any air in the steam that could cause boiler corrosion, and then to the boiler feed pump which raises the pressure to that of the boiler. A proportion of the condensate is analyzed for purity and chemically treated to maintain the quality and thus reduce boiler corrosion.

The efficiency of the cycle is improved by the use of feed-heaters and air heaters. Feedheaters are fed with surplus steam from tapping points on the turbines and preheat the condensate before it enters the boiler. The latent heat in this steam is thus not lost from this cycle. Similarly airheaters are located at the exhaust of the boiler at the entrance to the chimney, where the incoming air is preheated by the combustion gases.

The fuel is burnt with a controlled amount of air to ensure complete combustion and minimum pollution. Oil and natural gas are readily burnt in the furnace; coal must be ground to a fine powder, known as pulverized fuel, before it can be readily burnt. The burnt coal forms a very fine ash which is collected in electrostatic precipitators to avoid being deposited on the surrounding countryside. This ash (called pulverized fuel ash or pfa) is a useful by-product with wide applications from land reclamation to making lightweight building material.

A 2000 MW coal fired station burns 4 m tons of coal each year. It is transported direct from the pit by trains that can discharge the coal automatically through bottom hoppers while on the move. An equivalent oil station would burn 2½ m tons of oil; these stations are sited near refineries if possible for pipeline supply.

Cooling water in the condenser is contained within thousands of tubes and never comes in contact with the steam to avoid contamination. During the pass through the condenser the water picks up almost as much heat as is usefully produced by the cycle, such are the losses due to latent heat. The cooling water temperature rises 8 to 10°C (14 to 18°F) while passing through the condenser. Where the cooling water is drawn from a large body of water such as the sea, an estuary or a large river, it may be pumped directly back to the source and the heat is dispersed by mixing and natural cooling. If such a source is not available then the cooling water is recirculated through cooling towers and the heat dissipated to the atmosphere.

Nuclear power stations

In a nuclear power station the heat from the nuclear reaction is used to generate the steam. This steam then passes through the turbine, condenser and condensate system as in a conventional station. As the maintenance of nuclear reactors is complicated by radioactivity, there are more stringent quality controls on the water purity to ensure a minimum of boiler corrosion.

There are two main types of thermal reactors in commercial use and their particular characteristics have a direct effect on the design of the associated turbine. These two types are gas cooled and water cooled reactors.

Gas cooled reactors include the Magnox, Advanced Gas Cooled (AGR), and High Temperature (HTR) reactors. Both AGR and HTR operate at a sufficiently high temperature to generate steam suitable for modern conventional turbines and high cycle efficiencies are obtained.

The reactor core and boilers are located inside a prestressed concrete pressure vessel. The heat generated by the nuclear reaction is removed by coolant gas under very high pressure, which is then pumped to the boilers. The boilers have three sections (economizer, evaporator and superheater) but unlike the conventional boiler these sections are simply different parts of the same tube, a *once through* boiler. This arrangement is adopted to reduce the number of penetrations through the pressure vessel.

Water cooled reactors include the Boiling Water (BWR), Pressurized Water (PWR), *CANDU* and Steam Generating Heavy Water (SGHWR) reactors. The coolant must not boil within the reactor core as steam would absorb too many neutrons and thus kill the reaction. Therefore the water is kept under sufficient pressure to avoid boiling. The steel pressure vessel, despite wall thicknesses of 12 inches, cannot withstand the pressures used in conventional cycles so lower pressures and steam temperatures are adopted.

The turbine design associated with this plant differs from conventional turbines in that larger volumes of lower grade steam must be handled, requiring larger turbine rotors and casings. Higher stresses are placed on the rotating plant and a solution adopted for the largest units is a turbine that runs at half synchronous speed with a four pole alternator. Cycle efficiencies are lower than for gas cooled reactors, but the construction costs are lower.

Nuclear stations are more expensive to build than conventional stations, but their operating costs are less than half. The saving is achieved in the fuel cost. One ton of Magnox fuel generates as much energy as 15,000 tons of coal, and AGR or PWR fuel even more. As a safety precaution, early nuclear power stations were sited away from centres of population and the system loads. Coastal sites were chosen for the supplies of cooling water. Fuel transport to these remote sites was not a problem given the small quantities involved, for example one truck load for 660 MW AGR instead of 750 coal trains. With improved designs, particularly prestressed concrete pressure vessels, a relaxation of siting criteria is being introduced, enabling siting close to the loads and shorter transmission lines. Nuclear stations generate base loads, and high load factors have been achieved. The problem with nuclear power is disposing of the radioactive waste materials, which are a more series pollution problem than any other created by man. (See also *Nuclear reactor* on page 50.)

Hydro-electric power

Water is stored behind a dam to a given level, then allowed to be run off through culverts round the dam to water turbines in a hydro-electric power station at its foot. The vertical distance the water falls through from the surface to the turbine is called the head. The higher the head the greater the power that can be generated, and the larger the lake formed behind the dam the greater the energy available overall.

Hydro-electric stations are sited on rivers with a good fall or slope and high flow, or at the natural outflow to an existing lake which is fed from local water courses. The effect of the dam on a river is to create a lake behind it, and suitable sites are found where a good head can be produced without an attendant flooding of a large area; rivers flowing in canyons or gorges are well suited, the rock walls containing the lake. The water turbines rotate slowly, 100 to 300 rpm, and consequently have multi-pole alternators to generate at system frequency. Generation may be on base load, peak load or seasonal, depending on the availability of water throughout the year and the integration of the station within an irrigation or water control scheme.

Pumped storage

Pumped storage power stations are a prime example of energy storage. Such stations are found in mountainous areas where two lakes at markedly different heights may be connected by culverts. The water turbine and generator units in the station are mounted on common shafts with pumps. During the night, when the system loads are low, and there is surplus economic generating capacity available water is pumped from the lower lake to the upper one using the generator as a motor to provide the drive. During the day, when loads are high and less efficient and more expensive power stations are operating, the peak loads are met by running water down from the upper lake to the lower one through the turbine. Although the overall efficiency of the system is only 70%, there is a net saving from pumping with cheap electricity and generating when it is expensive. The plant is very expensive, but may be used to meet rapid changes in load as well as peak loads.

Gas turbines

Gas turbine generators range in size from 2 MW to 70 or 80 MW, with some as high as 175 MW. They use either purpose built industrial gas turbines or converted aircraft type engines. The principle of the gas turbine is as follows: air is drawn in and compressed, passed to the combustion chamber where fuel is injected and burnt, and the resulting combustion gases exhausted through a power turbine connected to the compressor. Drive to the alternator is either directly through a gearbox from the power turbine, or indirectly by exhausting the gases to a second power turbine connected to the alternator. These arrangements are necessary as the turbine rotates at a higher speed than the alternator. Industrialized versions of aircraft engines are modified to allow continuous operation at sea level, derated to extend time between overhauls and adjusted to burn industrial fuels such as gas oil, light virgin naphtha or natural gas. Typical power outputs are 12 MW for an Avon engine and 17 MW for an Olympus. Multiple arrangements of engines driving one generator are used, giving unit sizes up to 70 MW. Overhauls are at 2000 hour intervals by removing the engines to the central works. Industrial gas turbines are of more rugged construction, comparable to small steam turbines, and operate at lower temperatures than the aircraft type; 600 to 800°C (1112 to 1472°F) against 900 to 1100°C (1652 to 2012°F). Major overhaul periods are 80,000 hours.

Gas turbine generators have efficiencies of 20 to 26% and, burning refined fuels, are expensive to operate. Their construction costs are low, half those of conventional plant, and

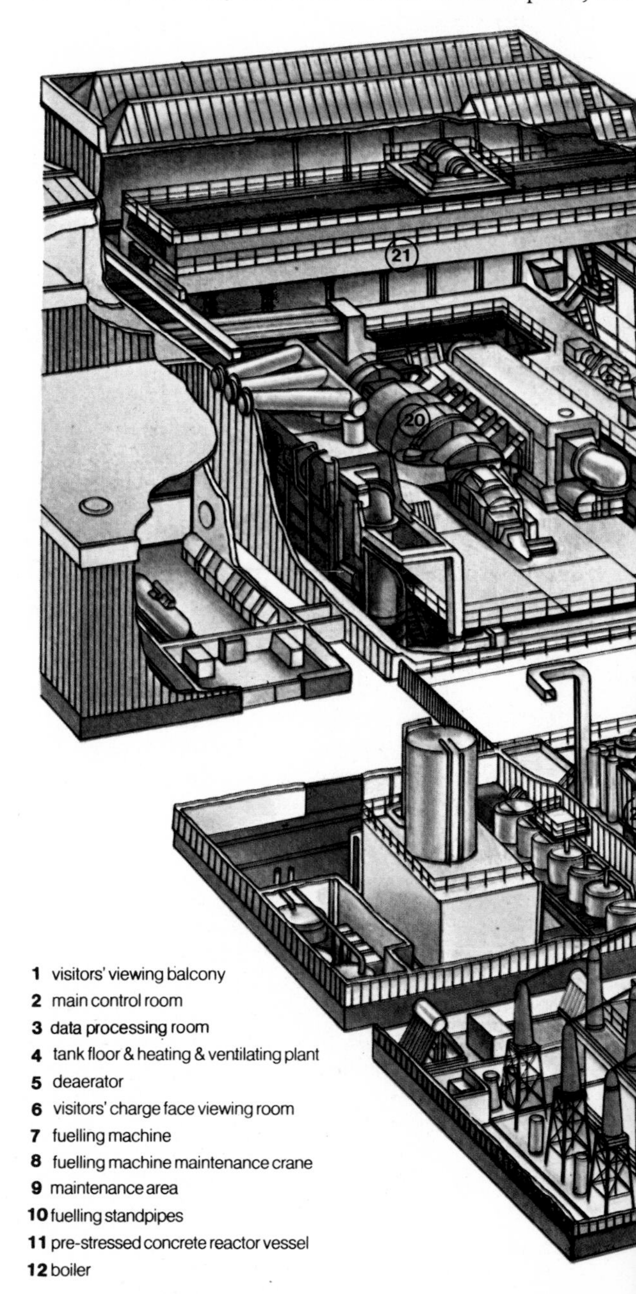

Below: an Advanced Gas-cooled Reactor nuclear power station. It uses more than 170 tons of slightly enriched uranium dioxide in each of its two reactors.

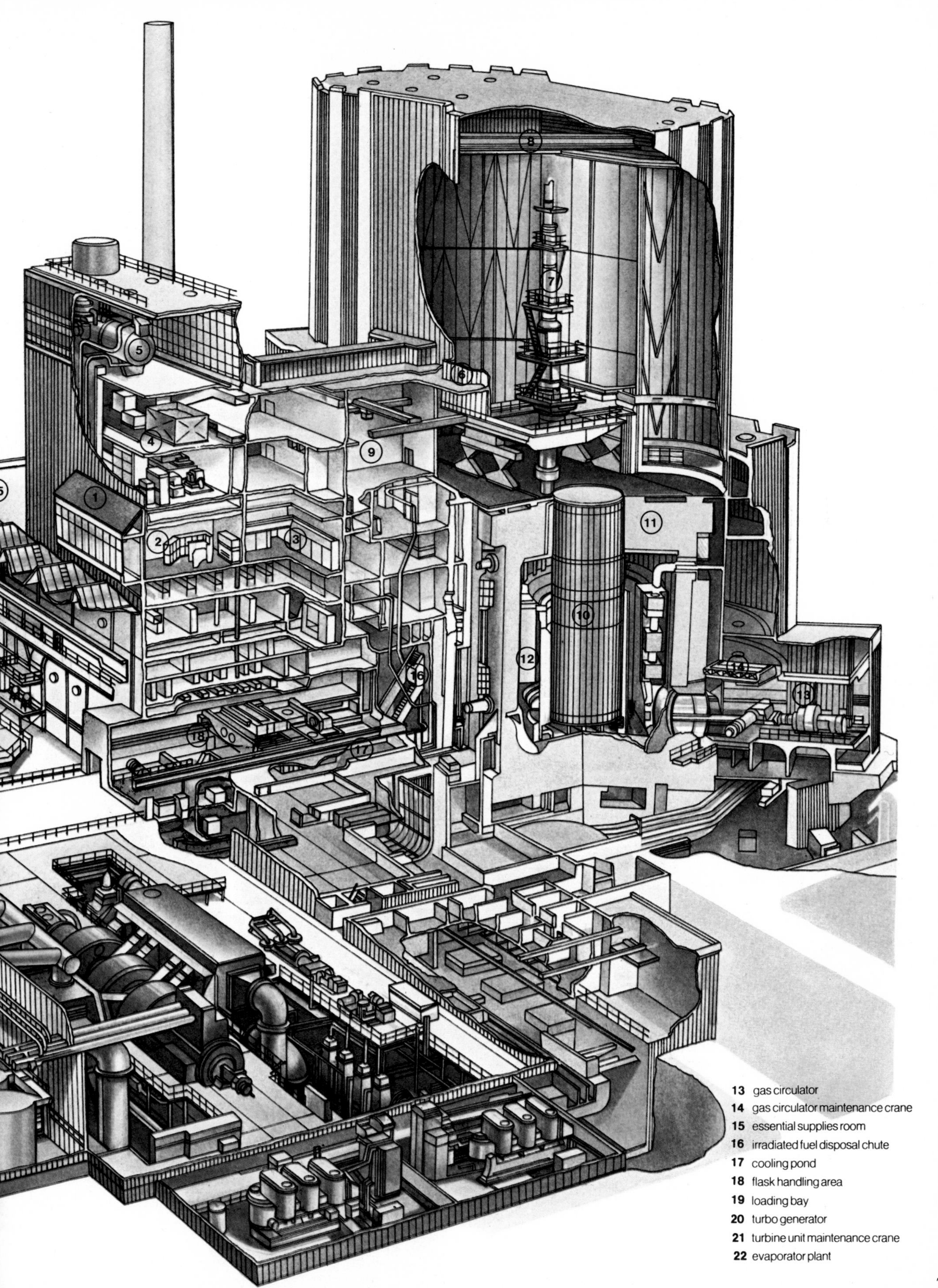

13 gas circulator
14 gas circulator maintenance crane
15 essential supplies room
16 irradiated fuel disposal chute
17 cooling pond
18 flask handling area
19 loading bay
20 turbo generator
21 turbine unit maintenance crane
22 evaporator plant

Right: a 330 kV cable termination under construction in the Snowy Mountain hydroelectric scheme, Australia.

with their quick loading they are suited for peak loads; annual load factors (average load divided by the maximum load) are of the order of 10%. Air is the working fluid of a gas turbine; five times as much air is drawn in than is required for combustion, giving 80% air in the exhaust. The high exhaust temperatures are the major losses of the cycle. Some of these losses are recovered in a combined cycle power station. In this, the waste heat in a gas turbine exhaust is recovered by passing the gas to a waste heat boiler where steam may be generated and passed to a steam turbine. The high air content of the exhaust permits further combustion of oil fuel in this boiler, raising the power output of the system. Typically a boiler giving 3 to 4 times the output of the gas turbine alone may be used with an overall efficiency of 35 to 40%. The system offers rapid start peak loading from the gas turbine and less flexible but more efficient operation from the steam plant. If the gas turbine is conservatively rated and cheap fuel available, the plant may approach base load operation on a small system.

Magnetohydrodynamics

Just as hydrodynamics deals with the way in which water and other fluids move, magnetohydrodynamics or MHD is the science of the way in which fluids which are good conductors of electricity move and interact with magnetic fields. The fluid can be either a liquid (such as mercury, molten sodium or brine) or a gaseous conductor, which is called a plasma. Gases can become ionized and therefore able to conduct electricity, under the influence of heat or another source of energy.

When any conductor moves through a magnetic field, an electromotive force (emf) is generated, which drives an electric current in a direction at right angles both to the field and to the direction of motion. A typical example is a dynamo with solid copper conductors, but the same effect applies with a fluid conductor. With a fluid, the forces can change the flow inside it so the details of the fluid movements must be considered in addition to the electromagnetic effects. It is this coupling together of the equations of motion with the electromagnetic equations that complicates MHD theory.

When dealing with gases, the subject is sometimes called *magnetogasdynamics* (MGD)—that is the Russian term—or *magnetoplasmadynamics* (MPD). The French use the term *magnetoaerodynamique*, but its unfortunate abbreviation, MAD, has prevented its use in English.

If a hot ionized gas is driven through a magnetic field, the current which is generated magnetohydrodynamically can be picked up by a pair of electrodes. In this way, power can be extracted directly from a high temperature flame without the need for all the usual intermediate stages—boiling water to raise steam to pass through a steam turbine to drive a shaft to turn a dynamo. As a direct generation process, MHD potentially has considerable advantages.

An important factor in MHD generation is the conductivity of the gas and this depends on the type of gas involved and its temperature. Typically, at a temperature of 3000°C the concentration of ionized atoms is only about one part in a million. At 4000°C this may have increased to

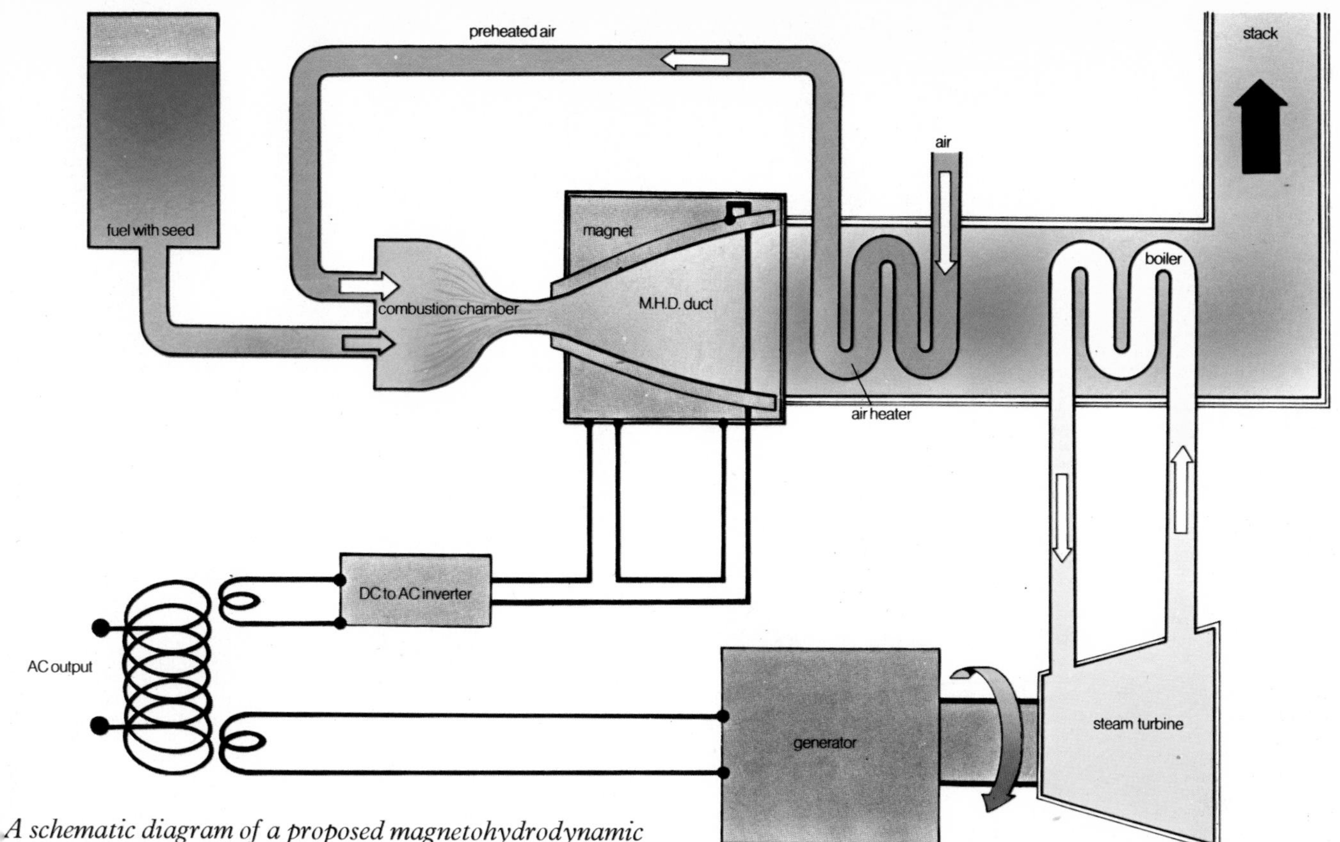

A schematic diagram of a proposed magnetohydrodynamic open cycle combustion plant with conventional steam turbine. The hot ionized gas (plasma) enters the chamber at high velocity and because it is ionized behaves like a fast moving conductor. The flux is tapped at right angles.

one part in 10,000, but this is still far too low for efficient conduction through the gas.

One way of increasing the ionization concentration is to 'seed' the gas with an element more easily ionized at the temperatures involved. For example, at 3000°C the addition of a percent or so of potassium to the gas increases the ionization concentration to about one part in 1000. The gas then has a conductivity of about 40 to 50 mhos per metre (that is, a resistivity of about 0.02 ohm-metre).

Unfortunately, even when the gas is seeded with some easily ionizable material such as potassium and it is raised to temperatures as high as 3000°C to 4000°C, its conductivity is still only a millionth of that of copper. The volume of an MHD generator is therefore much bigger than that of an equivalent copper wound dynamo, its internal resistance is much higher and its efficiency as a generator is lower.

When the power taken from the hot flame, however, has reduced the temperature to around 2000°C, the conductivity is too low for further MHD use but the gas still contains a great deal of heat and this can be used in a conventional generating station. So the MHD generator is a thermodynamic 'topping' device which can increase the total efficiency of conventional plant.

The MHD electrical power which is generated comes from slowing the gas down to extract kinetic energy and from expanding it to release compression energy. The hot gas leaves the combustion chamber at around its sound speed of 2250 mph (1000 m/s)—sound travels about three times faster at flame temperatures than it does at normal temperatures.

The very high flame temperatures can be obtained by burning the fuel (gas, oil or even powdered coal) in oxygen. For large scale use, however, it will be necessary to burn the fuel in air suitably preheated by the hot exhaust gases before they are cleaned, to recover all the seeding material, and released to the atmosphere.

The heat and friction losses at the electrodes and at the insulating side wall surfaces are enormous—conditions are very similar to those in the nozzle of a Saturn rocket—but the MHD power is generated from the whole volume. To produce more power than is lost, it is necessary to build large multimegawatt generators whose volumes are large in relation to their wall area.

Russia and the USA have each built MHD generators which burn oil or natural gas to produce tens of thousands of kilowatts of electricity for a minute or so, a few thousand kilowatts for a few hours and lesser powers for hundreds of hours; they aim to develop MHD power stations which will generate millions of kilowatts.

Liquid metal MHD systems are being developed. The heat from a nuclear reactor would boil a metal such as sodium and the vapour would force the liquid metal through an MHD duct. Although the conductivity of the liquid metal is very high, the bubbling and foaming of the boiling process causes great problems that have yet to be solved.

The electromagnet needed to provide a very high magnetic field over the large volume of MHD duct calls for hundreds of tons of copper and steel and many megawatts of electrical power. An alternative would be to use a very large superconducting magnet. The successful development of MHD generation will call for new materials and techniques to handle the very high and very low temperatures which can be produced.

Transmission of electricity

Electricity is not a primary source of energy, neither is it directly usable as it requires the consumers' apparatus to convert it to light, heat or motive power. Its usefulness lies

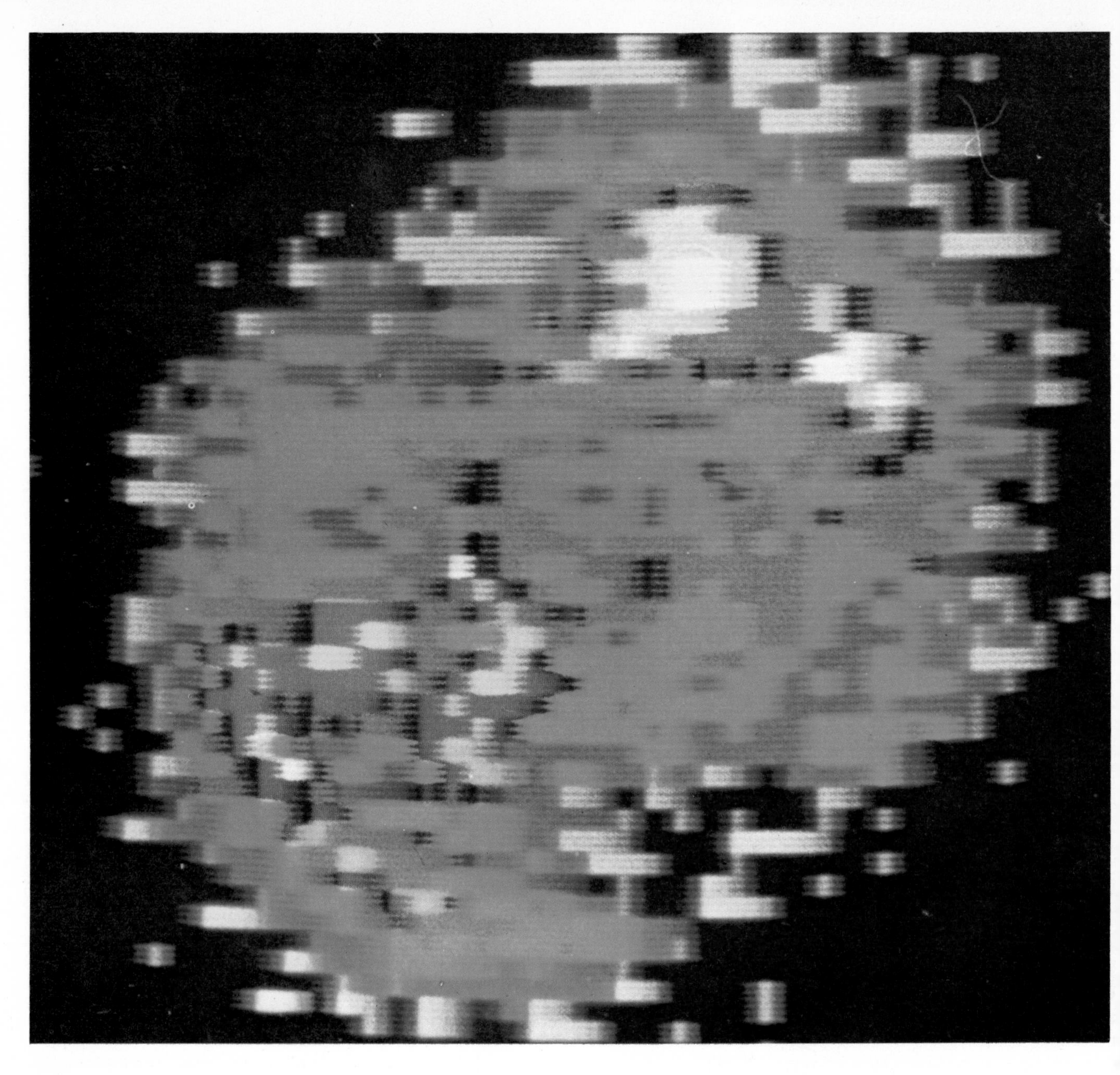

in the relative ease with which it can be manipulated and conveyed over long distances compared with other forms of energy.

To transfer energy from power station to consumer the electrical transmission system has been developed from the earliest local distribution systems, working at about 200 volts, to the modern national networks which generally operate at about 400 kilovolts.

Power losses

The basic circuit for transferring electrical power from source to load comprises two conductors, one providing the 'forward' path and the other the 'return' path, enabling the current to pass through the load and so do work. The function of the source is to provide a voltage or electrical pressure to establish and sustain the current in the circuit.

If the source provides a steady DC voltage, the magnitude of the current will be determined not only by the resistance of the load but also, marginally, by the resistance of the conductors and the source, in accordance with Ohm's law. The total power in the circuit will be the product of the voltage and the current, or the sum of the power delivered to the load and the circuit power losses.

Above: spectroheliograph in digital form (by satellite) of solar storm activity. MHD can be studied in the motion of solar plasma. White regions are areas of intense storms while yellow and red clusters denote lesser intensities.

The power losses in the conductors are proportional both to their resistance and to the square of the current, that is, if the conductor cross-sectional area is halved (doubling the resistance) then the losses are doubled, and if the current is doubled, the losses are increased fourfold. As power is equal to current times voltage, for the same power doubling the voltage enables the current to be halved, thus quartering the losses and improving transmission efficiency fourfold.

In an AC circuit, the current and voltage waveforms are sinusoidal and their effective values are 0.707 times their peak values. Theoretically an AC two wire system requires twice the conductor material as an equivalent DC circuit to transmit the same power to a resistive load at the same efficiency.

A further phenomenon associated with AC circuits is that the circuit components, particularly reactive (capa-

Below: diagram showing the way in which power is brought from the power station to the consumer, and typical voltage levels at each stage.

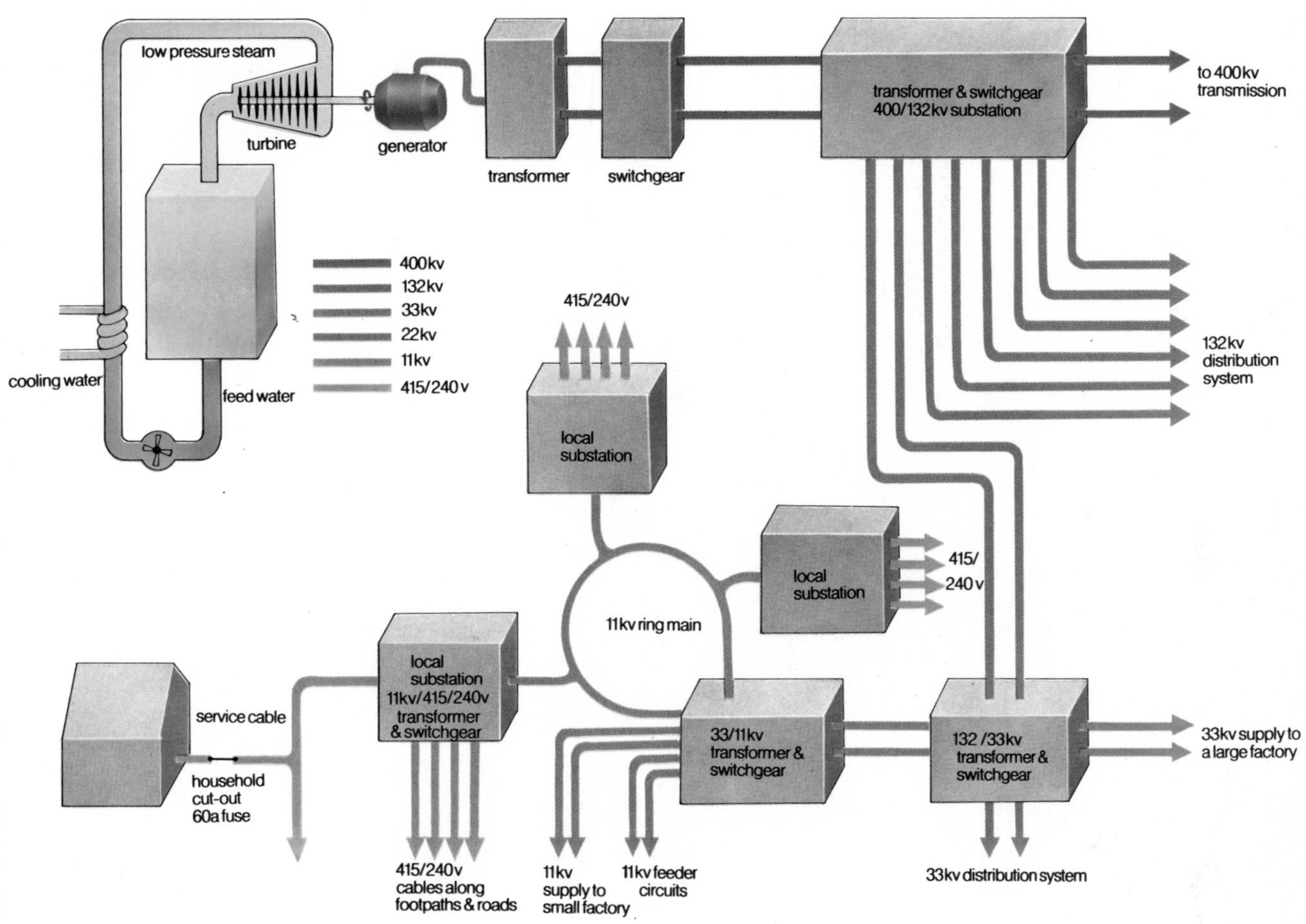

citive or inductive) loads, such as electric motors, exhibit additional characteristics due to the current-creating magnetic and electrostatic fields which temporarily store and then release energy throughout the cycle. The effect is to displace in time the voltage and current waveforms from each other, and the effective power in the circuit is reduced by a power factor which is proportional to this displacement. As a result of this phenomenon, some current flows in the circuit without doing any useful work.

The conductors of the supply cable must be insulated both from each other and from their surroundings, and the insulation must be able to withstand the maximum voltage of the system, which under abnormal transient conditions may rise to a value substantially higher than the maximum working voltage. In practice the system is deliberately connected to earth (ground) at the neutral point of the source to render the insulation requirements more predictable under all conditions and achieve economies in insulation material.

AC and DC

Direct current (DC) electricity, such as is supplied by a battery, always flows steadily in the same direction along a wire. Alternating current (AC) electricity, however, which is used for mains supplies in most countries, flows first in one direction then the other, reversing its flow at a certain frequency — usually 50 or 60 cycles a second (written 50 or 60 Hz).

For a 50 Hz frequency, the current builds up to a maximum in one direction and drops to zero in the first hundredth of a second. It builds up to a peak in the opposite direction and drops to zero again in the next hundredth of a second, making a fiftieth of a second for the entire cycle.

A light bulb or an electric heating element works equally well whichever way the current is flowing, and so do electric motors designed for AC operation. So the fact that the current oscillates is quite unimportant in practice.

The advantages of AC rather than DC power supply is that its voltage can be stepped up or down using transformers which have no moving parts. AC motors and alternators (AC generators) have no commutators — divided metal slip-rings for picking up the current — and are thus more reliable than their DC counterparts, which do have commutators. The frequency is chosen as a compromise between the conflicting requirements of transformers, power lines, lighting, rotating machinery and so forth.

The standard AC power systems adopted throughout the world are 3-phase 3-wire and 3-phase 4-wire. The 3-wire system is used mainly for high voltage distribution, the 4-wire system being found predominantly in low voltage distribution, where both 3-phase and single phase loads have to be supplied. In certain circumstances it may be preferable to use a high voltage DC link for the bulk trans-

Left: a 750 MVA, 400/275 kV 3-phase transformer at a large grid substation. The station connects a higher voltage station to a lower one; power can be more efficiently transmitted over long distances at voltages too high for local use. Hence the necessity of the substation transformer. Opposite page: work on a live 11kV line. The equipment is mainly made of glass fibre and the ropes are nylon. 'Live' working means more continuity of service for the consumer of electricity; nowadays even damaged poles can be replaced while the circuit is operating.

mission of power, in order to avoid certain technical limitations created by electrostatic and magnetic effects associated with high voltage alternating current.

Distribution

The voltages used in electricity supply systems vary somewhat from one country to another, but in general the basic distribution principles are fairly similar. The following description is a general account of a typical modern system.

Large modern generating sets generate electricity at about 22 kV and are typically grouped in stations each having a total capacity of 2000 MW or more. This voltage is too low for the economic transmission of large quantities of power over long distances, and so the voltage is raised to 400 kV by means of a transformer connected to each generator. The transformers are linked via high voltage circuit breakers to the switching station from which overhead or underground cables connect to the power system's high voltage network.

Frequently the power station is remotely sited and the high voltage system conveys power efficiently over long distances to tapping points where there are large demands for it. At these points, a substation is established, comprising more high voltage switchgear, transformers to step down the voltage (to 110 kV or 132 kV), and a lower voltage switching station. To this switching station are connected the lines and cables of the lower voltage system.

If the substation is required to serve more than one township, the low voltage system may consist of several overhead lines to carry the power to the small and medium sized towns in the area. If it is within or close to a large city, the lower voltage system may be comparatively short underground cable circuits.

The power is transformed successively via numerous distribution networks (from 10 to 30 kV), which individually handle power levels appropriate to the demand in the areas they serve, until it reaches the low voltage system (240 or 110 volts single phase), to which ordinary domestic consumers are connected. Large consumers, such as factories, are connected at one of the intermediate voltages; the greater the power required the higher the voltage they are connected at.

It would be impossible to supply all loads on a single circuit, and domestic consumers usually have a single circuit supply. As the size and importance of the load increases, however, it is considered necessary to have at least a duplicate supply so that if a fault develops, or if maintenance to equipment is required, power can still be supplied to the consumers. This gives a high degree of security of supply.

Cables and overhead lines

Where possible, electricity is transmitted by overhead lines, which are cheaper and technically simpler than underground cables. For these reasons power is carried by overhead lines in rural areas, but in urban areas it is frequently impossible to obtain a route for an overhead line and the increased costs and technical problems of underground cables have to be faced.

At the lower voltages (about 30 kV and below) employed in distribution systems, cost differentials are much smaller and the technical difficulties do not arise. Underground cables are therefore common at these voltages even in quite small towns.

Substations

The purpose of a substation is to connect a higher voltage network to a lower voltage network by means of transformers, and the main items of equipment at a substation are switchgear and transformers.

Switchgear consists of busbars, circuit breakers and isolators. A circuit breaker must be capable of interrupting the abnormally heavy currents which flow in a short circuit fault and of closing or opening a circuit on load. An isolator, except at the lower voltage, is an off load device used to isolate a circuit (or part of a circuit) or is a selector switch to connect a circuit breaker to one busbar or another. An isolator is a simple device, basically consisting of a fixed contact and a moving blade. The busbars are used to connect together all the circuits in a switching station, and are made of copper bar or tubing mounted on ceramic insulators.

Protection

All systems are liable to faults whether caused by natural factors such as a salty atmosphere in coastal regions or by lightning storms, or by accidental damage such as a crane jib striking an overhead line. Usually one conductor is short circuited to another conductor or to earth (ground). Under these circumstances the circuit must be de-energized as quickly as possible and this is done automatically by protection equipment.

The simplest protection of all is the fuse, which both detects and interrupts a fault when the heavy current melts the fuse wire. Fuses are used at voltages up to about 11 kV. Where circuit breakers are installed, other protective devices detect the faulty current and send a signal to the tripping coil of the circuit breaker, which then opens to interrupt the flow of current in the faulty circuit.

In some cases the protection operates when a pre-set current level is exceeded. The protection can incorporate a time delay or be almost instantaneous in operation. If several sets of this type of protection are installed to protect various sections of a circuit, the ones at the remote end would operate very quickly and those nearer to the main substation would have progressively longer time delays. This ensures that a fault at the end of the circuit results in only that part of the circuit being switched off. This is suitable for radial circuits, supplied from one end only, and confined to voltages up to about 11 kV.

More complex systems measure the current entering and leaving a circuit and operate only when there is a difference in these values. Thus faulty current flowing right through a healthy circuit into a faulty circuit will not cause the healthy circuit to be opened.

Another system measures both current and voltage to determine the distance from the circuit breaker to the fault. It will not operate immediately if the fault lies beyond the circuit it is protecting.

Synchronism

In general, power systems are 3-phase AC operating on a fequency of 50 or 60 Hz, and this means that when a generator is run up to speed to go on load it must be switched into the system only when it is running in synchronism with the other generators in the system.

The voltage of each phase of the generator must reach its peak at the time as the corresponding phase of the rest of the system, otherwise unacceptable fluctuations would occur until the generator was 'tripped' out again. To synchronize the generator, its speed is adjusted very slightly up or down as necessary until the voltage peaks match. At that precise moment, the circuit breaker controlling the generator is closed and the set is now synchronized to the system.

Stabilization

The stability of a power system is its ability to withstand abnormal disturbances without violent fluctuations in voltage, frequency or loss of synchronism. Over the length of a circuit, due to the effects of induced magnetism in the conductors (reactance), the voltage peaks gradually become displaced relative to the sending end voltage. The power transmitted along that circuit is determined by a mathematical function known as the power-angle rule. As the angle between the voltage peaks increases, the power transmitted increases until it reaches a maximum at an angle of 90°, and then decreases.

ELECTROMAGNETIC

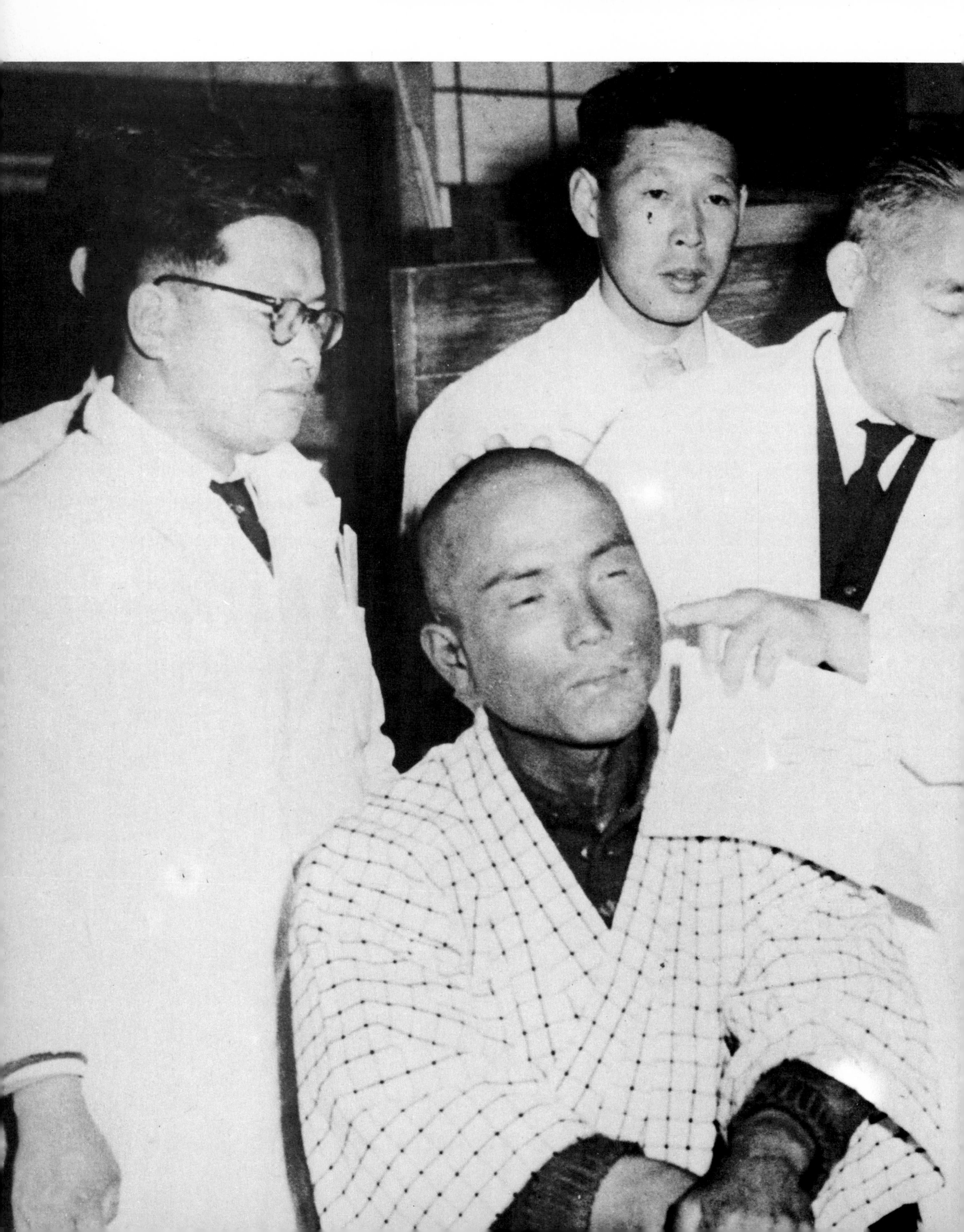

RADIATION

Japanese fishermen were burned in 1954 by radioactive ash from an atomic test at Bikini. They were outside the supposed danger area.

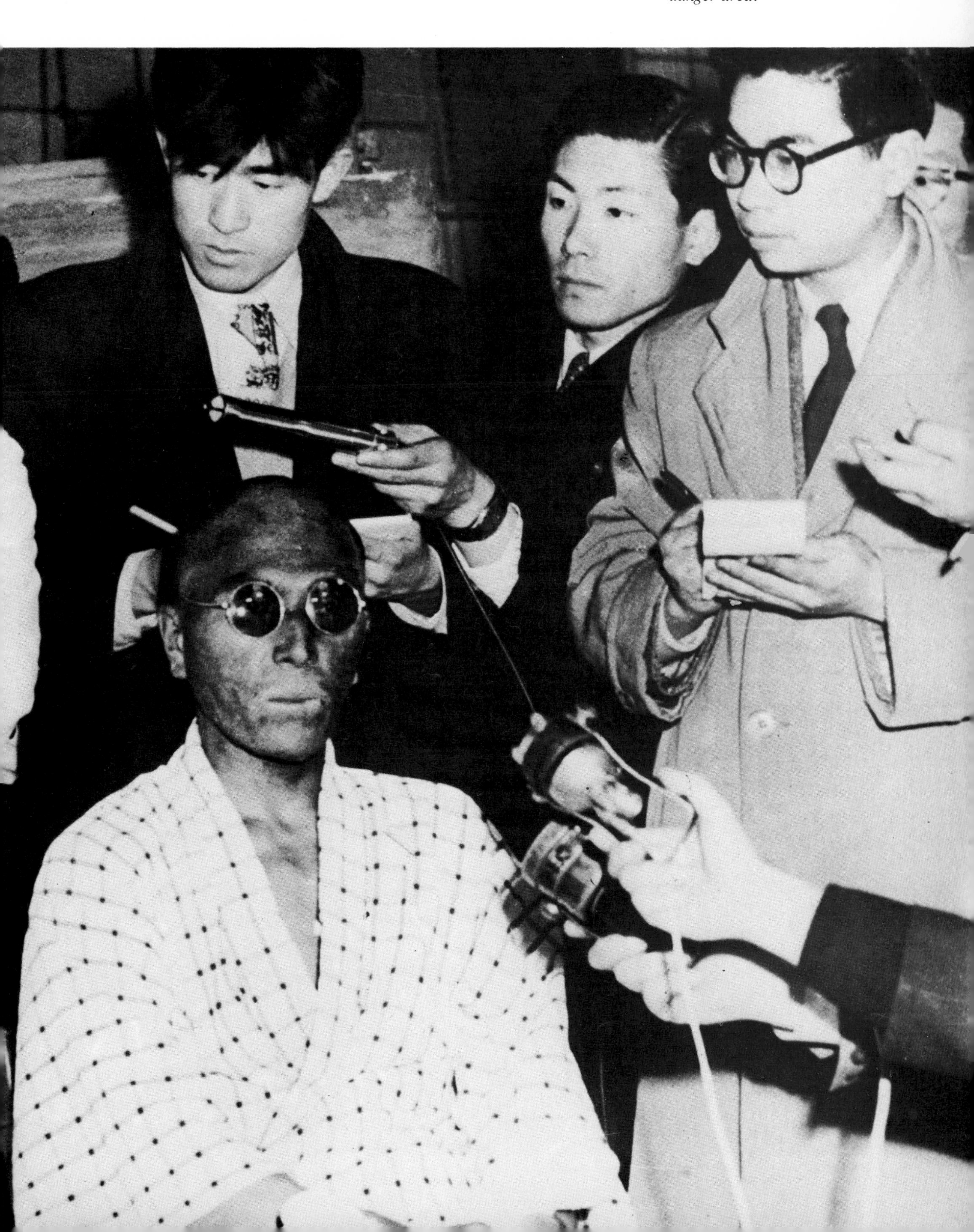

Many apparently unrelated phenomena, such as light, radi waves and X-rays, are different examples of just one type o radiation, electromagnetic radiation. They are in fac waves of energy produced when an electric charge i accelerated.

A stationary electric charge—in practice, a charge particle such as an electron—is surrounded by lines o force which indicate the direction in which another simila charge would move if it were placed near the origina charge. If the charge is moved up and down it is decelerated and then accelerated in the opposite direction a each end of the path. These accelerations cause 'kinks' i the lines of force, which move outwards from the charge A moving charge, however, also generates a magnetic field (This is the principle of electromagnetism, in which th current of moving electrons produces a controllabl magnetic field). An accelerating charge generates a 'kinked magnetic field, whose lines of force are perpendicular to th electric field. The speed at which these 'kinks' move outwards depends on what kind of material is surrounding th charge, but in a vacuum it is 299,792.456 km per secon (186,282 miles per second).

Left: ripples on sand are caused by the wind, just like waves on the sea. The advanced mathematics necessary to explain wave formation was not developed until the 1950s.

James Clerk Maxwell was the first to calculate that these 'electromagnetic' disturbances could exist, and would travel at this velocity. He noticed that this is the same as the measured speed of light, and suggested that light is a form of electromagnetic radiation. At the time it seemed necessary that these 'waves' should be vibrating in some mysterious medium, which was called ether. This had the curious properties that it filled all of space, was far more rigid than steel, yet could not be detected except by the vibrations in it. Einstein's theory of relativity, however, showed that electromagnetic waves can travel without any medium to support them, and so the ether is no longer believed in.

Frequency and wavelength

To produce a continuous wave of electromagnetic radiation, the charge must be vibrated up and down continuously. The number of vibrations of the charge in one second is called the *frequency* of the resulting wave, and is measured in cycles per second or *Hertz* (Hz), after the scientist who first produced and detected radio waves. The lowest frequencies of interest are around 150,000 Hz (150 kHz), which are 'long wave' radio frequencies. VHF radio uses radiation at about 100,000,000 Hz (100 MHz), while light is at very much higher frequencies (600,000 GHz) and X-rays higher still (3,000,000,000 GHz).

Another way of distinguishing types of electromagnetic radiation is by the wavelength, the distance between successive 'crests' of the wave. For any type of wave it must be true that velocity of wave = frequency × wavelength.

Electromagnetic radiation travels at different speeds in different materials, so the wavelength varies according to the medium, but the frequency is always constant. The 'wavelength' of a particular radiation usually means the wavelength it would have in a vacuum. For example, the yellow light emitted by a sodium lamp has a wavelength of 589.3 nanometres in a vacuum. (1 nanometre is one thousand millionth of a metre, and is abbreviated to nm. Wavelengths can also be given in angstroms, where 589.3 nm = 5893 A). In air it is reduced to 589.1 nm, and in glass it is only 388.6 nm.

The longest radio waves are more than 10,000 m (over 6 miles), while the shortest waves (gamma rays) are at wavelengths less than 0.001 nm, smaller than an atom. (Note that low frequencies correspond to long wavelengths, and high frequencies to short wavelengths).

Below left: theKyrenia Ridge in Cyprus induces downwind waves in the air; clouds occur at the 'crest', where the air is cooler. Below: the electromagnetic spectrum. Different detectors are needed for each section. The superimposed 'black body' spectrum shows wavelengths emitted by the Sun (6000°C) and an electric bar heater (2000°C).

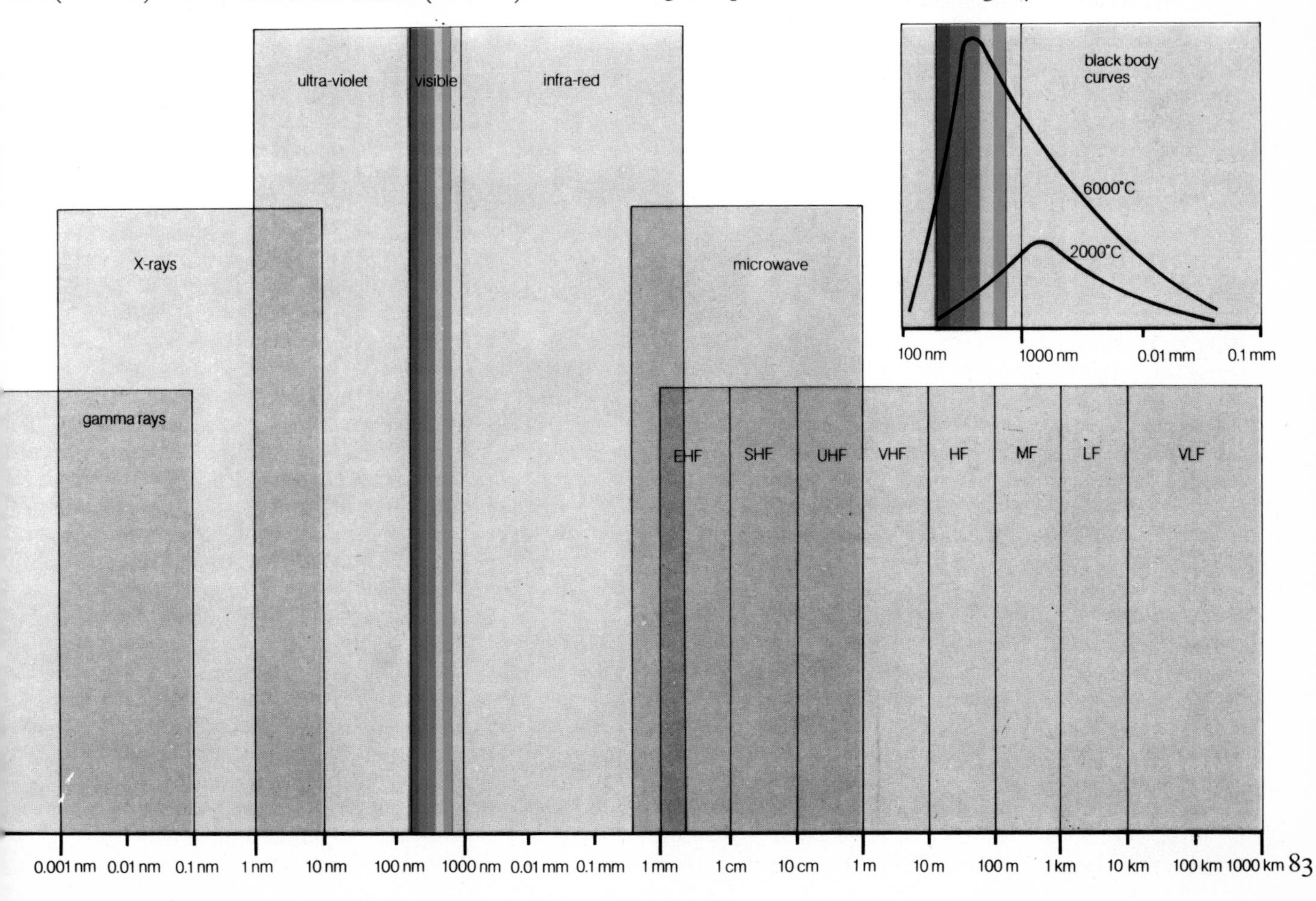

RADIO WAVES

Above: a row of microwave aerials at the Post Office relay station, Ingleby Arncliffe, Yorkshire. The microwave system is used for messages of national importance; the narrow beams are aimed precisely, making it hard to eavesdrop.

Opposite page top: sky and sunrise colours are caused by scattering of light in the atmosphere. Sunlight is white, but air molecules scatter the shorter wavelengths, so the sky and distant mountains appear blue. The direct light is more red, but this is only obvious when the Sun is low.

Electromagnetic waves longer than 1 mm are known as radio waves. They are subdivided into groups, such as very high frequency (VHF) and ultra high frequency (UHF), depending on their frequency. The very longest waves usually detectable are called VLF, for very low frequency, with wavelengths longer than 10 km and frequencies lower than 30,000 Hz. At this end of the scale, it becomes rather impractical to detect the signals, which are of very low energy.

Radio transmitters work on the principle of rapidly switching on and off an electric current. Whenever current is switched on, as in, for example, some domestic appliance, one pulse of electromagnetic radiation—one 'kink' in the lines of force—is produced. If the current is switched on and off at a high frequency, then electromagnetic radiation of that frequency will be produced. This is how radio transmitters work in principle: electrons are forced to pulsate at the chosen frequency along the transmitting aerial (antenna), which should for greatest efficiency be as long as the wavelength being emitted.

The electrons which constitute the current are accelerated as the current changes, and they radiate electromagnetic waves whose electric field is parallel to the transmitting aerial. If this aerial is vertical, only a vertical aerial will receive the radiation; it is said to be vertically *polarized*. Similarly a horizontal aerial will radiate horizontally polarized radiation, which can be detected only by a horizontal receiving aerial. (Polarization can be produced in all forms of electromagnetic radiation, and has useful properties.) The electromagnetic wave produces currents in the receiving aerial which are amplified in the receiver to reproduce the transmitted message.

The highest frequency that can be produced electronically is about 300,000,000,000 Hz, corresponding to a wavelength of 1 mm. Since the term 'radio' usually refers to electronically produced radiation, this marks the end of the radio region.

Higher frequencies can be reached by using the natural vibrations of the molecules in a solid. These molecules contain electrons which generate electromagnetic radiation as the molecules vibrate. The hotter the solid is, the more rapidly the molecules vibrate, and the higher the frequency of the radiated electromagnetic waves. Radiation produced in this way is normally unpolarized, because polarization due to electrons moving in different directions will tend to cancel out.

LIGHT

The phenomenon of light, the explanation of what it is, and of why objects are visible at all, is not easily understood even today. The earliest known explanation of vision was by the sixth century BC Greek philosopher Pythagoras, who hypothesized that an object is visible because light rays travel outwards from the eye to touch it. This theory failed to explain why we cannot see in the dark, and so the idea was elaborated by Plato (fifth century BC), who required in addition 'emanations' from the Sun and from the object observed. Within the following couple of centuries, however, the true explanation (that light is emitted from luminous bodies like the Sun or a flame, and is reflected off other objects into our eyes) had been widely accepted.

The nature of light rays was not satisfactorily explained until the present century. At the time of Sir Isaac Newton there was heated controversy between those who (like Newton) believed light to consist of rapidly moving particles (or *corpuscles*), and the followers of the Dutch physicist Huygens who regarded light as being a series of waves. The latter theory was strongly supported by the diffraction experiments of Thomas Young in 1801, whose results could not be explained in terms of the corpuscular theory, and also by Maxwell's theoretical prediction of electromagnetic radiation (consisting of oscillating electric and magnetic fields moving as a wave) whose properties were identical to those of light. The quantum theory, introduced at the beginning of the twentieth century, showed that when light is emitted or absorbed it behaves like a particle (known as a photon). The energy of a photon depends inversely on the wavelength of the associated electromagnetic wave—that is, short wavelengths are most energetic, and long wavelengths the least. The real nature of light is not easy to imagine, and it is simplest to regard it as a wave-particle duality, with the two aspects being important in different circumstances. Einstein's theory of relativity also did away with the necessity for a medium to transmit the radiation.

Light of only one wavelength is termed monochromatic, and it is usually incoherent; that is, the light consists of short wavetrains, or strings of waves, each lasting about one hundred millionth of a second, with the peaks and troughs of each wavetrain occurring randomly relative to every other wavetrain. The light of a laser owes its intensity partly to the fact that the peaks of all the wavetrains are in phase: it is *coherent*.

Most of the objects we see are not self-luminous, but are only visible by reflected light. If the surface has irregularities larger than the wavelength of light the reflection is diffuse, the light being scattered back in all directions. At a very smooth surface, however, light is reflected specularly. It is reflected back in such a way that the angle the ray makes with the perpendicular to the surface is equal to that made by the incident ray. These conditions make possible the formation of images.

If the body does not strongly absorb or reflect light, but is transparent, the light will pass through it. Its velocity is different in different media, and consequently the ray changes direction at the surface. This phenomenon, refraction, affects each wavelength to a different extent, and so the incident light is spread out according to wavelength. The formation of this spectrum of colours (corresponding to different wavelengths) was first investigated by Sir Isaac Newton in 1665. A spectrum can also be produced by a very finely ruled diffraction grating.

INFRA-RED RADIATIO

Two hundred years ago the spectrum of what we now call electromagnetic radiation was thought to comprise only the visible colours from violet to red. But in 1800 Sir William Herschel discovered that the Sun radiates energy beyond the red end of the visible spectrum: he made the first measurements at infra-red wavelengths. What we now call the infra-red occupies a far larger section of the spectrum than Herschel could have anticipated, bridging the gap between visible light and microwaves. Only in the last few decades has man learnt to make use of infra-red radiation: it is of great value in such divergent fields as medicine, cookery, the study of gases and military technology.

Herschel's detection of infra-red radiation was made in a neat experiment which can be easily repeated. He passed sunlight through a slit and then a prism to produce a spectrum on a table top. Into the various colours of the spectrum he inserted the blackened bulbs of thermometers. The blackening absorbed all colours equally, so the temperature to which the mercury rose indicated the amount of energy carried by the light of each colour. The Sun emits most of its energy in the yellow-green; because a prism spreads out blue, green and yellow light and bunches up the red, however, Herschel's thermometers read highest in red light. This prompted him to sample beyond the red and so to make his discovery.

The radiation Herschel discovered had a wavelength of

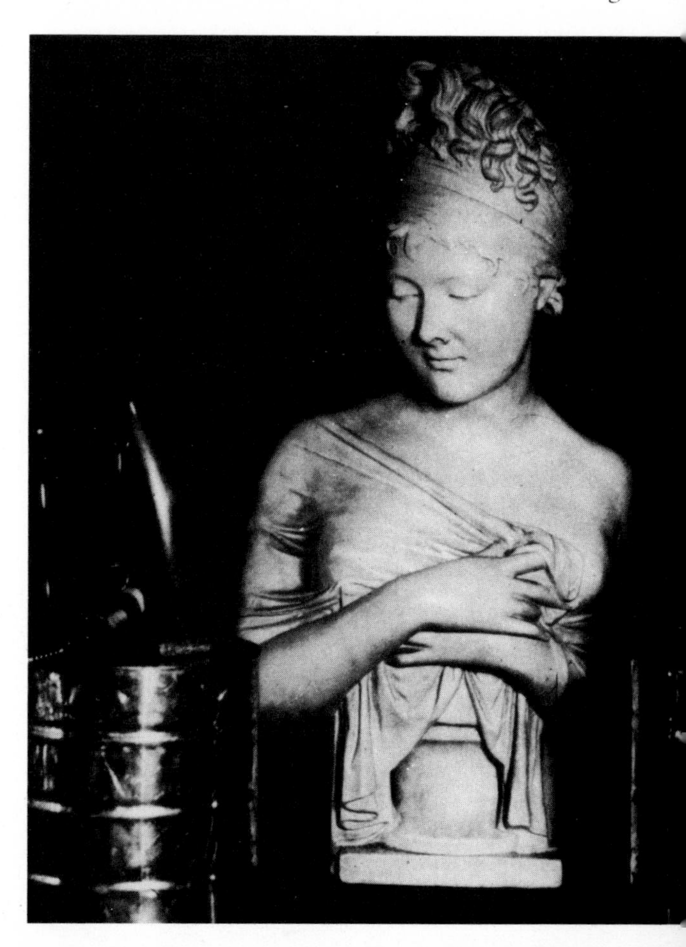

Velocity of light

The velocity at which light travels in a vacuum is fundamental to modern physics. According to relativity theory it is identical for all observers, whatever their own velocity, and it is also the fastest speed at which any material object can travel. The first measurement of the velocity of light (usually denoted 'c') was made in 1676 by Ole Romer, who observed that the movements of the satellites of Jupiter seemed to differ from the predictions because the time taken for light to reach the Earth varies as Jupiter and the Earth move around their orbits.

Until the development of electronics after World War II the most accurate measurements of c were made by timing a light ray over an accurately measured distance, as in Fizeau's method. At the far end a mirror reflected the light back to the observer, who measured the very short time taken, by means of a rapidly rotating toothed wheel. The outgoing light is chopped into bursts by the teeth, and at a critical wheel rotation velocity the teeth will have advanced far enough to cut off these bursts when they are reflected back. At this wheel velocity the reflected light is seen to become very dim, and the time taken for the light to make the return journey can be calculated from rotation velocity of the wheel and the number of teeth on it. Later and more accurate experiments (by Michelson and others) used revolving mirrors instead of toothed wheels, and enclosed the entire light path in a vacuum chamber to minimize the effects of air.

The most accurate measurements of c have been made recently using radar techniques to measure separately the frequency and the wavelength of microwaves, another form of electromagnetic radiation, and give the result c= 299,792.458 km/second (186,282 miles/sec).

about 0.8 micrometres (millionths of a metre, abbreviated μm and usually called microns). Visible light covers the range 0.4 to 0.7 microns and the infra-red extends to 1000 microns (1 millimetre). Infra-red radiation is invisible to human eyes but can be sensed by our skin just as Herschel's thermometers detected it—by its heating properties. When an electric fire is switched on we can feel its radiant heat, due to infra-red, before the element begins to glow red. Throughout the visible spectrum, the cooler an object is the redder it appears. This relation applies in the infra-red. Objects cooler than about 900°C emit essentially all their energy in the infra-red. Room temperatures bodies emit most strongly at a wavelength of 10 microns.

Detecting infra-red

If our eyes were sensitive to 10 microns radiation we should not need artificial lighting: everything would be bright day and night. Living objects would stand out clearly being warmer, and hence brighter, than their surroundings. Cold objects would appear dark: we should have difficulty finding things inside a refrigerator. The ability to 'see' warm objects even in the dark is of obvious military value, and has promoted much research into detection systems superior to thermometers. Most infra-red detectors are of the single element type: they do not produce two-dimensional pictures of a scene but merely measure the total amount of radiation falling on them. The most versatile of these detectors are the bolometers which respond to radiation of any infra-red wavelength.

Infra-red detectors are used on guided missiles to direct the weapons at a warm target. Infra-red imaging devices—*vidicons*—have recently been developed: these respond to 10 micron radiation and display a room-temperature scene on a television screen. Some snakes have 10 micron 'vision' which enables them to catch their prey by night.

Photographic emulsions can be made sensitive to light no longer than 1.1 microns wavelength—the near infra-red. These wavelengths are considerably longer than the diameters of air and dust molecules, so a photograph taken on infra-red film will penetrate haze and mist much better than one taken on blue sensitive film: blue light is easily scattered by the air. A certain type of infra-red colour film is available in which the colours of objects appear shifted in the spectrum—blue light does not show up at all, green objects appear blue, red objects appear green and infra-red objects appear red. This 'false colour' film has proved useful in spotting from the air areas of vegetation with different infra-red reflectivities.

Uses

Infra-red radiation can be employed either to transmit heat to an object or to detect radiant heat from it. It heats directly by radiation, rather than by conduction, so it can heat objects quickly; an infra-red grill will cook a steak in

Opposite page top: halos are rainbows are caused by the diffraction of light by water droplets.
Far left: the photograph was taken with normal light. Then (near left) another photograph was taken with infra-red film with the lights turned off. The electric irons illuminated the scene for the infra-red film, although nothing was visible to the naked eye.

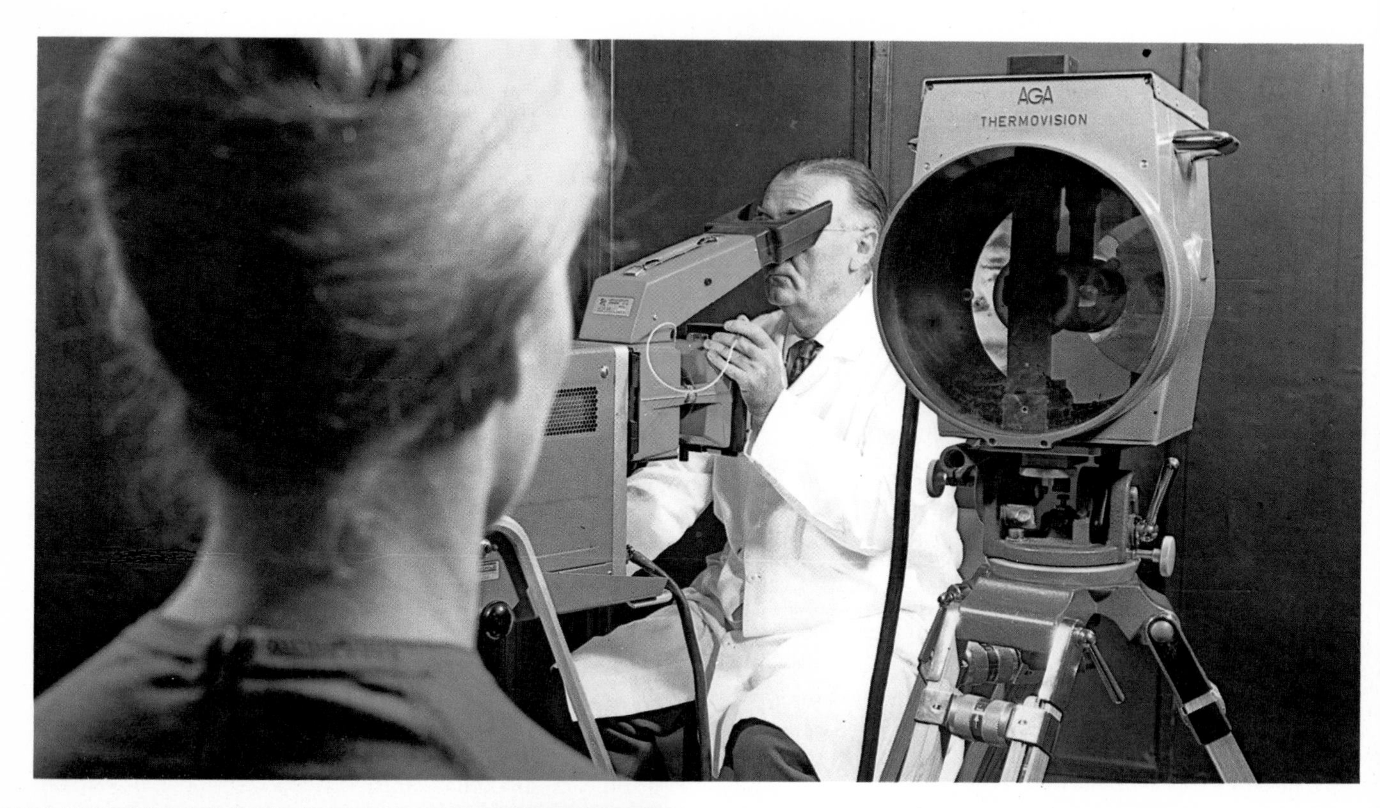

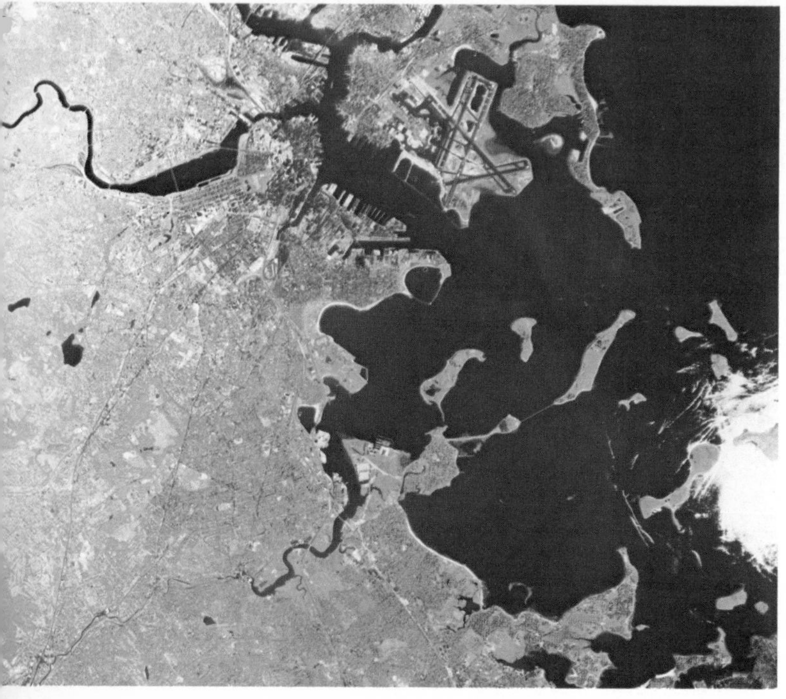

two minutes.

Just as human skin is slightly transparent to red light, it is even more transparent to light of longer wavelength. Infra-red can therefore penetrate to some depth, and infra-red lamps are used by physiotherapists in the heat treatment of muscles and tissues. Alternatively, ten micron vidicons can be used to study the temperature of human skin, a process called *thermography*. Thermographs indicate areas of the body where the blood flow is abnormal, and are helpful in diagnosis.

Gaseous molecules have natural frequencies of vibration which occur in the infra-red, so each type of molecule absorbs infra-red radiation in different wavebands. Spectra of gases are therefore like fingerprints: by mapping the transmission at different wavelengths in a spectrophotometer the composition of a sample of gas may be deduced.

Water vapour and carbon dioxide in the Earth's atmosphere absorb infra-red radiation at all but a narrow selection of wavebands, the atmospheric windows. Conveniently, one of these occurs at 10 microns and allows distant viewing of room-temperature scenes. We can also examine objects outside our atmosphere. Infra-red astronomy has led to the discovery of extensive clouds of dust in nebulae and around certain stars; some of these may represent the birth of stars and planetary systems.

Above: aerial view of Boston, Massachusetts, using infra-red false colour film. Vegetation reflects infra-red even more strongly than green, hence the red overall colour. Film records only a small part of the infra-red spectrum, called the 'near' infra-red.

Top of page: taking a thermograph picture of a person's head. A reflecting system, like that of a telescope, is used rather than a lens, because a lens tends to absorb light, while a mirror reflects all colours equally. The infra-red diagnostic technique of thermography is useful because it records relative temperatures.

ULTRA-VIOLET

At wavelengths shorter than 390 nm is the ultra-violet, which extends down to 1nm. This radiation is emitted by extremely hot bodies, but the temperatures needed are higher than the boiling point of all substances so ultra-violet is produced this way only in very hot stars.

On Earth ultra-violet is produced in a different way. The electrons in atoms and molecules can have only certain energies, and when they move from one energy state to another they emit the excess energy as electromagnetic radiation. This radiation will occur at particular frequencies corresponding to the energy changes in the atom. Many atoms will produce frequencies which are in the ultra-violet part of the spectrum, one example being mercury, which is used in 'sun-tan' ultra-violet lamps. Atoms can also produce wavelengths which lie in the visible spectrum by this process. The colour of sodium street lights is due to an energy change in the sodium atom which results in radiation whose wavelength corresponds to yellow light.

Above: minerals viewed by white light (top) and ultra-violet light. The lower four are clearly the same mineral (flourite) although their normal appearance differs.

X-RAYS

The penetrating radiation emitted when a beam of electrons strikes the glass wall of a discharge tube was discovered accidentally by W. C. Röntgen in 1895. He named it 'X-radiation', as its nature was at that time completely unknown. It is now recognized that X-rays are electromagnetic radiation of wavelength between 0.001 and 10 nanometres, considerably shorter than the wavelength of light (about 500 nm).

The medical uses of X-rays are probably their best-known application, but the penetrating power of X-radiation is also important in industry for inspecting materials and welds for internal flaws. Further important applications occur in metallurgy and geochemistry, where X-rays are used to analyze both the chemical composition and the crystalline structure of specimens.

The earliest X-ray sources were developed from Röntgen's gas discharge tube, but in these it was difficult to keep a constant gas pressure, and consequently the tubes had only a short life. In 1913 Coolidge developed the type of X-ray tube in general use today. The glass tube is highly evacuated and the electrons are produced at the cathode (negative electrode) by an electrically heated filament similar to that in a thermionic valve (vacuum tube). The electrons are accelerated towards the positive electrode (anode, or anti-cathode) by the electric field, and they strike a 'target' of tungsten set into the copper anti-cathode. Here 1% of the energy of the electron beam is converted into X-radiation, which escapes through a 'window' in the otherwise well-shielded tube. The other 99% of the energy is converted to heat, and so the copper anti-cathode must be water cooled while the tube is in operation.

In the Coolidge X-ray tube the intensity of the X-radiation is controlled by the current in the cathode filament, while an increase in the voltage between cathode and anti-cathode allows shorter wavelengths to be generated at the target. Generally speaking, shorter wavelength X-rays are less easily absorbed and are therefore more penetrating. Hospital X-ray machines usually operate at voltages less than 150,000 volts, although Coolidge tubes can be used up to several hundred thousand volts, while special types incorporating several electrodes operate up to 2 million volts. X-radiation of even shorter wavelength (less than 0.001 nm) can be produced by letting the electrons in a particle accelerator strike a metal target. These short X-rays are physically identical to gamma rays; the latter name is used when the radiation is of radioactive origin, while it is termed X-radiation if it is produced by the sudden stopping of high energy electrons.

The impact of the electrons on the target causes some X-ray emission just because of the sudden deceleration of the electrons when they collide with free electrons in the target material. This *bremsstrahlung* (a German word meaning 'braking radiation') is, however, very much less intense than the radiation from the atoms of the target. The latter occurs when an electron in one of the innermost orbits of the atom is 'knocked out' of the atom by a high energy electron from the cathode. An electron from the next lowest orbit immediately takes the place of the missing

Below: a ceiling-mounted X-ray tube with infinitely variable positions.

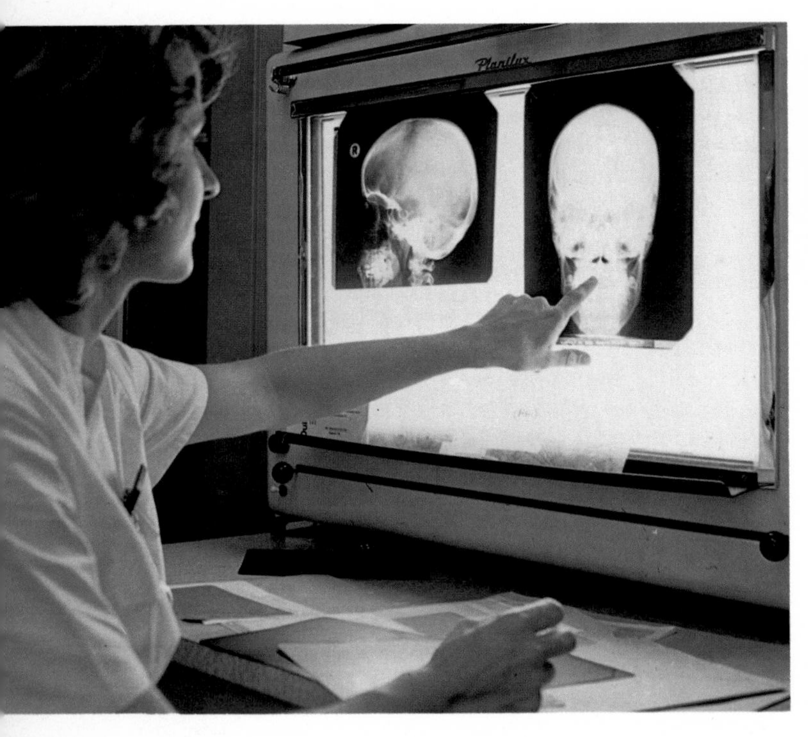

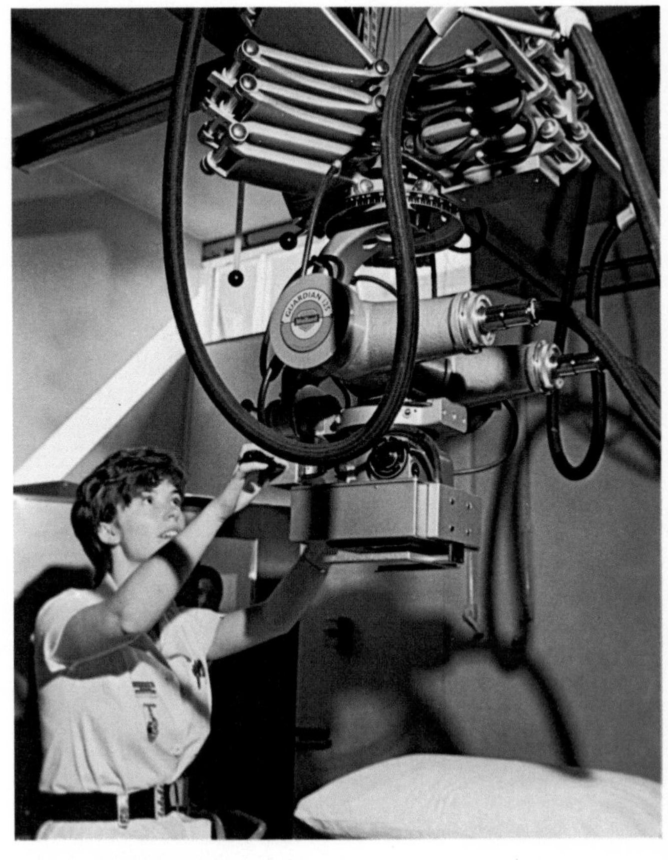

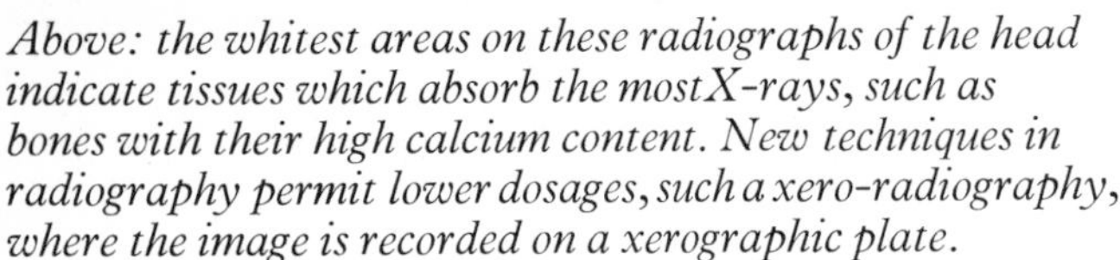

Above: the whitest areas on these radiographs of the head indicate tissues which absorb the mostX-rays, such as bones with their high calcium content. New techniques in radiography permit lower dosages, such a xero-radiography, where the image is recorded on a xerographic plate.

electron and this change of orbit causes energy to be lost as electromagnetic radiation of a particular wavelength (line emission). For the inner electrons of atoms with a high atomic number, this radiation is at X-ray wavelength. Each element has a characteristic X-ray spectrum, just as it has a characteristic optical spectrum due to changes in the orbits of the outermost electrons. Monochromatic, or single wavelength, X-radiation can be produced by filtering out all the other X-ray lines from the target by means of suitable absorbing filters.

Radiography

X-rays are absorbed at different rates by different materials: for X-rays of a particular wavelength the absorption depends on the atomic number (equal to the number of electrons) of the atoms present, and on the density. In radiography the sample is placed between the X-ray source and a photographic plate; the more absorbing parts of the specimen throw shadows on to the plate, and these appear less dark when the plate is developed. (If the photographic plate is replaced by a fluorescent screen the image can be seen directly; this technique is called fluoroscopy). A radiograph of the human body, for example, shows the bones whiter than surrounding flesh because bones contain the element calcium, which has a relatively high atomic number. Any abnormalities, such as broken bones, arthritis, or foreign bodies in the stomach or lungs are readily visible, and appropriate therapeutic action can be taken. Internal organs generally absorb X-rays to about the same extent as the surrounding flesh, but they can be shown up on a radiograph by concentrating material of greater absorbing power into the organ. A 'meal' of the insoluble salt barium sulphate shows up the stomach and intestinal tract, while iodine compounds injected into the blood-stream become concentrated in organs such as the thyroid gland and the kidneys.

Radiography also has important industrial uses in locating internal defects in materials, particularly for inspecting welds. The movement of internal parts of machinery can be followed by motion picture radiography; exposures of a hundredth to a thousandth of a second are obtained from an X-ray tube with a rotating target which lies within the electron beam for only this brief time. Another specialized type of radiography used in both industry and medicine is stereoscopy, where two radiographs are taken with the X-ray source moved by the distance between the human eyes. When these are viewed through a stereoscope a three-dimensional image is seen, and the position of, say, a bullet in the body can be instantly detected.

An important new technique for studying the three-dimensional structure of solids is tomography, originally developed to investigate brain tumours. The X-ray source, the patient and the photographic plate are moved continuously during the exposure in such a way that the shadow of one particular plane through the brain is always in the same position on the plate, but the shadows of the

Below: using an X-ray source to calibrate a silicon photon spectrometer.

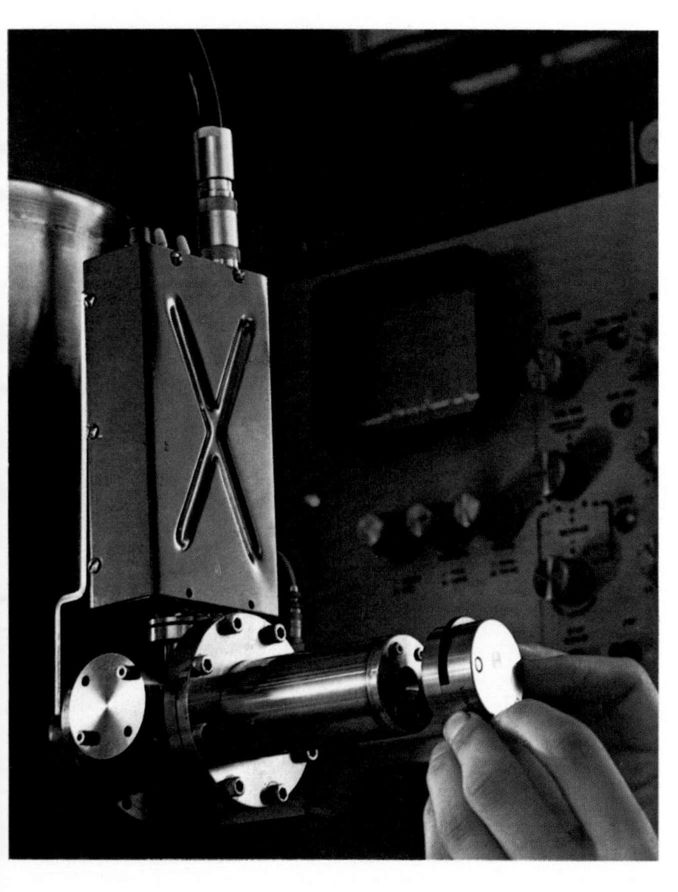

rest of the brain are in continuous motion and are thus completely blurred out. The result is a radiograph showing a cross-section through the tumour, and from many such sections in parallel planes a detailed three-dimensional picture of the tumour can be built up.

Radiology is a branch of medicine which also includes both radiography and the therapeutic uses of radiation. X-rays are often used to destroy cancerous growths, although great care must be exercised as the very high doses required can also damage healthy tissue; in extreme cases an overdose can produce leukaemia (cancer of the blood cells).

X-ray astronomy

Natural sources of X-rays include the Sun, especially around the dark sunspots, and certain double stars where matter from a giant star falls on to a dwarf star and is heated to a temperature of millions of degrees. At these high temperatures, X-ray bremsstrahlung occurs when electrons collide with atomic nuclei, and this radiation has been detected by satellites above the Earth's atmosphere, which absorbs X-rays before they reach the ground. The dwarf star is sometimes a neutron star, with a density a million million times that of water; one particular X-ray star may even be a *black hole*, a region of space from which neither matter nor radiation (such as visible light) can escape. As matter spirals down into the black hole it is heated up and emits X-radiation before it disappears completely.

GAMMA RADIATION

The discovery of radioactivity by Henri Becquerel in 1896 inspired Pierre and Marie Curie to investigate the 'radiation' emitted by radioactive atomic nuclei. They discovered alpha and beta 'rays', which were later shown to be helium nuclei and very fast electrons respectively; but it was not until 1900 that Paul Villard found that a third type of radiation was also emitted. This had no electric charge, and could penetrate a block of iron a foot thick. It was called gamma radiation, and was later shown to be a form of electromagnetic radiation. The wavelengths of gamma rays are the shortest in the electromagnetic spectrum (less than 0.01 nm, or 0.1 angstrom), and so their energy is even greater than that of X-rays.

Gamma radiation is produced by the rearrangement of protons and neutrons in the atomic nucleus. In the nuclei of radioactive elements a neutron will occasionally 'decay' into a proton, an electron and a neutrino. The last two particles escape from the nucleus (the electron being called a beta particle), but the nucleus is left with a surplus of energy which it emits as a gamma ray photon.

Gamma rays are usually detected by a geiger counter, in which the high energy radiation removes some of the electrons from the atoms to which they are bound (a process known as ionization). The current of electrons is amplified to give a signal which can be recorded. An alternative technique is the use of a photographic emulsion sensitive to gamma rays, often called a 'nuclear emulsion'.

Ionization will also occur in the body tissues if these are exposed to alpha, beta or gamma rays, and, if the radiation is sufficiently intense, this can cause radiation burns, cancer, and death. Gamma radiation is particularly dangerous as it can penetrate the body to reach the vital internal organs. Workers who are likely to be exposed to gamma rays wear film badges consisting of a photographic emulsion in a light-proof cover which monitors the amount of radiation they are exposed to. The maximum safe dosage of radiation depends on the organ concerned; for example, the arms and feet are able to take fifteen times as much as the blood-forming organs.

Gamma rays have recently been detected from space, originating from both the centre of our Galaxy and the Crab nebula (the remnant of a star which exploded in AD 1054); there are also bursts of gamma rays whose origin has not yet been ascertained.

There are many beneficial uses of gamma rays. In medicine they are used to study disorders of the brain, thyroid, kidney, liver and pancreas. These organs will preferentially absorb very small quantities of a gamma ray emitting isotope administered to the patient and can thus be photographed by a gamma ray camera located outside the body. Concentrated gamma radiation is also used in medicine to destroy cancerous tissue in the body. Gamma rays are widely used in industry to examine castings and welds for flaws. A gamma ray source is positioned in front of the object to be examined and an image is recorded on a photographic plate located behind the object. In such applications gamma rays are generally used in preference to X-rays because X-ray machines are much larger and

DIFFRACTION

more expensive than comparable gamma ray equipment.

Hermetically sealed packs of food or surgical supplies are sterilized by intense gamma radiation which kills the potentially harmful bacteria, and the packaging prevents reinfection. The plastics industry uses gamma radiation to strengthen polythene and other polymers, because it creates cross-linkages between the long chain molecules which form the plastic. The properties of the plastic can be varied at will by changing the amount of radiation.

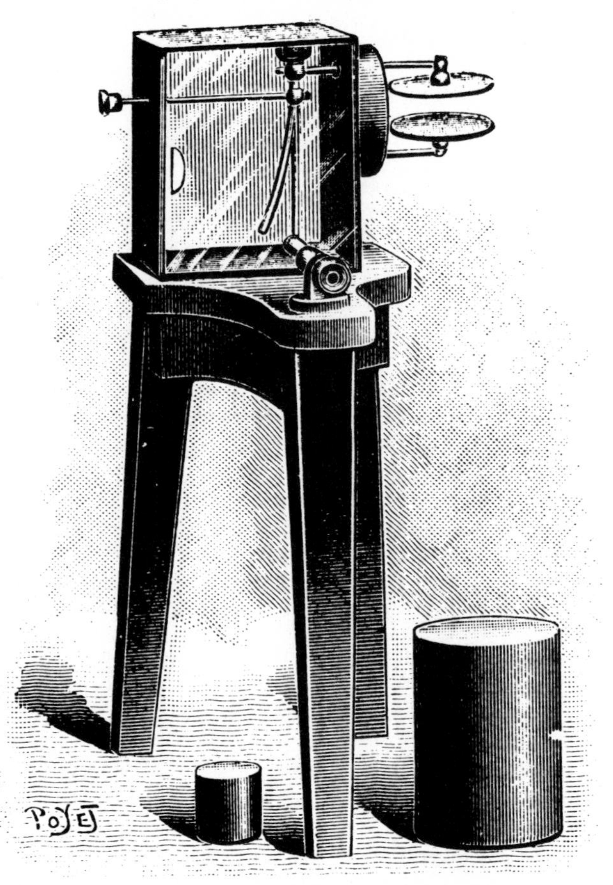

Above: the apparatus used by the Curies for detecting the presence of radioactivity. It consists of a gold leaf electroscope fitted with a microscope to detect the smallest movement of the leaf. The radiation from radium, for example, travels at very high speeds, leaving ionized atoms in its wake. A radium sample held near the electrodes of the detector causes charges to be accumulated on the leaf, which moves through electrostatic repulsion.

A seventeenth century Italian Jesuit, Francesco Grimaldi, was the first to discover that light does not travel in exact straight lines, but can bend round obstacles very slightly; this is caused by the phenomenon of diffraction. The effect is put to practical use to analyze light and crystal structure.

Grimaldi found that the shadows of thin rods illuminated by narrow shafts of light were not sharp, as would be expected. Instead there were bright lines just inside the edge of the shadow. The effect is not normally noticeable unless very small, bright sources of light or fine meshes are used.

If the filament of a clear light bulb or a candle is seen through a fine mesh, for example, the image will be seen to be broken up, with a series of images getting fainter with their distance from the main image. These images may be coloured, being red at one edge and blue the other. This breaking up of the image gives diffraction its name—it comes from the same Latin root as 'fracture'. The same effect is seen when a light is reflected off a fine-grooved record—delicate colours can be seen. Television camera-men use a grid over their camera lenses to produce a starlike effect on bright lights in variety programmes, and the spikes seen on photographs of stars are caused by support vanes inside the telescope used.

Diffraction can be observed in all types of wave motion: sound and water waves, as well as light, X-rays and radio waves, show the effects. It is because of diffraction, for example, that sound can be heard round corners. Practical observation of sounds heard in this way reveals another feature of diffraction: that long wavelengths are diffracted more than short ones. In the case of a marching band, it is noticeable that the notes which are heard first are those of the bass drum, while the piccolos are not clear until they are directly in view—even though the piccolos may then sound louder. Deep notes have a long wavelength whereas high notes have a short wavelength.

The same effect occurs with light, and gives rise to the coloured fringes associated with diffraction. As with sound, the low frequencies—that is, red light—are bent most, and the high frequencies (blue) are bent least. To produce the effect in a usable form, not just one 'corner' but a large number is required. A diffraction grating consists of a regular series of fine lines ruled on glass, which will either transmit light or, if given a suitable coating, reflect it. The spacing of the lines varies but in a high quality grating may, be as many as 30,000 per inch (12,000 lines/cm).

Dispersion

A narrow beam of light falling on such a grating will have bands of colour on either side of the direct image. There are several 'orders' or separate bands: as they get further from the direct image, they become more spread but dispersed and fainter.

The angle through which a particular colour is diffracted depends on its wavelength and the spacing of the grating's lines. The finer the grating, the more each wavelength will be diffracted and hence the greater the dispersion. A grating intended to diffract microwaves (with wavelengths several

Left: a diffraction grating splits up white light passing through a slit. This grating is aluminized to make it reflect; some of the light which comes through the slit is reflected as from a normal mirror, and some is diffracted into colours. The grating has some 30,000 grooves per inch (1200 per mm) and is 'blazed'; that is, the grooves are cut on an angle in such a way that the diffracted light is thrown only to one side of the white light, rather than to both sides. First and second order diffraction can be seen, the second being fainter.
Below: the spikes on the stars in this photograph are caused by diffraction around support vanes in the telescope; the rings, however, are a photographic effect.

nillion times longer than light) would therefore have a pacing of centimetres rather than thousandths of a centinetre.

X-rays, on the other hand, need much smaller spacings—hey have wavelengths thousands of times shorter than ight. The regular spacings of the atoms in crystals will iffract X-rays.

The study of crystal lattices using the X-ray diffraction atterns is known as crystallography, and techniques of this sort have proved invaluable in studying the properties of crystal structure. Electron and neutron diffraction are also used for this purpose, using the wave properties of matter.

Diffraction gratings are widely used in spectroscopy, the analysis of the wavelengths, whereas prisms, which split light into colours by refraction, spread the short wavelengths out more than the red end. Since the light does not have to pass through the glass, gratings can be used to study wavelengths which would otherwise be absorbed, such as infra-red and ultra-violet.

Although diffraction gratings of high quality have to be carefully ruled on optically flat glass, and are consequently very expensive, replica gratings of plastic materials are much cheaper. The replica is impressed with the fine lines by a high quality master die, in much the same way that a microgroove record is made.

Diffraction can be explained by using the model of wave behaviour first proposed by Huygens. He suggested that every point on an advancing wavefront can be thought of as a secondary source of waves in its own right. If undisturbed, these secondary waves will combine to form the new wavefront, so advancing it.

When an obstruction is placed in the way, however, some of the secondary waves spread out from the edge of the obstacle, so apparently bending round it. If a series of obstacles is used, these secondary waves may mutually interfere, either reinforcing or destroying each other's effect, so producing light and dark bands of light which are split into colours depending on the dispersion of each wavelength.

QUANTUM THEORY

Towards the end of the 19th century, scientists were well pleased with the neat, consistent picture of physics which they had built up. In particular, the theory of the nature and production of electromagnetic radiation was in a highly satisfactory and elegant state, James Clerk Maxwell having recently shown that the ultimate source of radiation is an accelerated electric charge which emits waves travelling at the speed of light. But problems began to arise when scientists considered the radiation produced by a class of objects called black bodies, which have the property of totally absorbing all radiation they receive. Several experimenters had measured the radiation emitted from black bodies, and there were a number of attempts to fit a mathematical form to the way in which it was distributed among different wavelengths. None of these attempts was completely successful, and it was rather disturbing to many physicists that when Lord Rayleigh used rigorous classical physics in his efforts to find a solution, he failed as badly as the rest.

The quantum

In 1900, the young German physicist Max Planck provided the answer, but only after he had been forced to conclude that the classical approach was not valid in this case. He daringly proposed that energy is radiated not as a continuous flow, but in 'bundles' which he called quanta (singular, quantum). The energy carried by each quantum is proportional to the frequency of the radiation, so that

$$\text{Energy} = \text{Frequency} \times \text{Constant}$$

The constant, written h, became known as *Planck's* constant.

Planck assumed that his quanta of radiation spread out like waves on the surface of a pond after leaving their source, allowing the wavelike behaviour of radiation to be explained.

The photoelectric effect

Experiments soon confirmed the quantized nature of radiation, especially light. Hallwachs had noted that some metals lost negative charge, later found to be electrons, when exposed to ultra-violet light, which has a higher frequency (and therefore higher energy) than visible light. As the intensity of the light increased, more electrons left the metal, but it was observed that there existed a certain limiting 'threshold frequency' of the incident light, below which no electrons were emitted.

These facts were very difficult to account for on the classical wave theory of light, but in 1905, Einstein, in his first fundamental piece of work, was able to explain them in terms of the newly-developed quantum theory. He extended Planck's ideas by proposing that quanta of radiation (called photons in the case of light quanta) do not spread out like waves after leaving the source, but maintain their character as discrete bundles of energy. Thus, the high energy photons of ultra-violet light are able to eject electrons from the surface of a metal by bombardment. An increase in the light intensity simply means that a large number of photons hit the surface and consequently eject more electrons, but these photons must be sufficiently energetic. If the frequency of the light falls below the threshold frequency, the photons will not have enough energy to eject any electrons.

The Bohr atom

The new concept of quantized radiation paved a way to the interpretation of the dark lines observed in the spectrum of the Sun. These lines result from the presence of certain elements, particularly hydrogen and iron, in the Sun's atmosphere, and although they had been discovered and mapped more than a century before by Fraunhofer and others, there was no ready explanation of them. Niels Bohr, a young Danish physicist working at Cambridge, combined the new model of the atom proposed in 1911 by Rutherford with the Planck-Einstein quantum theory, and in 1913 was able to account satisfactorily for the spectral lines of the hydrogen atom. Bohr's model retained the massive, positively charged nucleus and diffuse cloud of electrons of Rutherford's theory; its novelty lay in his non-classical assumptions.

Firstly, Bohr proposed that the electrons travel in orbits around the nucleus, but he suggested that only certain orbits were possible, each having a specific amount of energy associated with it. These are called energy states or energy levels. He secondly maintained that no energy is

Right: Niels Bohr with the particle accelerator at the Institute of Theoretical Physics, Copenhagen, of which he was director. He received the Nobel Prize in physics for 1922.

adiated or absorbed by an electron unless it 'jumps' etween two of these energy levels. The energy of the adiation then emitted (or absorbed) in the form of a hoton is equal to the energy difference between the two evels. In this way, Bohr avoided a problem of classical hysics; that is, a normally orbiting electron should emit adiation by Maxwell's laws, lose energy and spiral into the ucleus. Bohr was then able to explain spectral lines as eing due to radiation emitted or absorbed when an electron umps between two energy levels.

Bohr's theory was refined by Sommerfeld to allow for lliptical electron orbits, but even then it was difficult to pply it to atoms which were more complicated than ydrogen. But Bohr had paved the way for the far-reaching onceptual advances which were to put the quantum theory n a firm mathematical basis ten years later.

uantum mechanics

n 1924, Prince Louis de Broglie, working from the postu-ates of quantum theory, proposed that matter must also ossess wavelike properties, and that Bohr's model of the tom could be more easily explained by considering the lectron as a wave. The following year, Schrödinger, in ürich, developed this concept into the mathematical orm of wave mechanics in which he demonstrated that uantization of energy occurs naturally, and without ecourse to artificial assumptions. Simultaneously Heisen-erg, in Göttingen, introduced the technique of quantum echanics, having assumed that energy is quantized, and howed that the wave nature of matter followed as a natural consequence. It was a triumph for the theory of wave-particle equivalence that these two scientists, working independently, and from two opposite approaches, were able to produce complementary solutions.

eft: a pulse of laser light, photographed in a hundred housand millionth of a second as it passed through water. elow: according to quantum theory, particles of matter ught to have wave-like properties. An electron beam eflected by a nickel crystal is diffracted into zones.

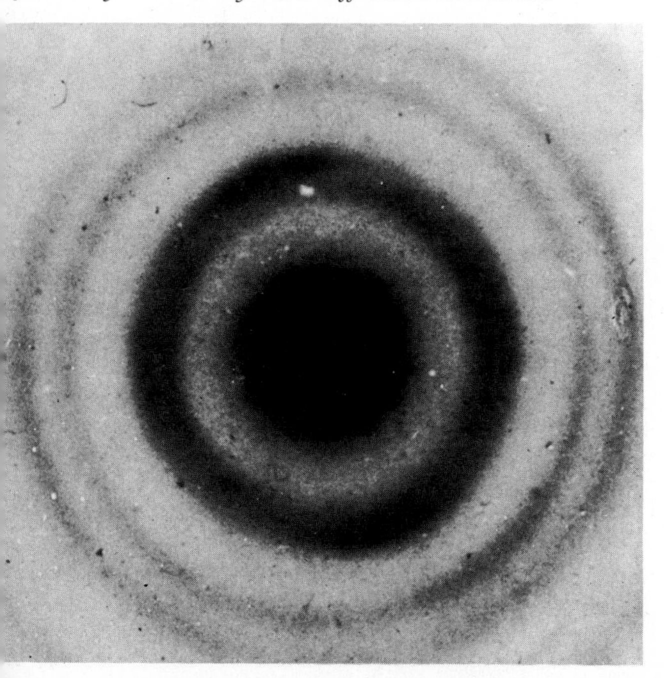

Heisenberg's work led to the formulation of his 'uncer-tainty principle' in 1926, in which he demonstrated that the wave-particle nature of matter makes it impossible to measure simultaneously and absolutely accurately both the position and momentum of a body. This restriction on measurement is fundamental, and is of philosophical as well as scientific importance. The effects of the uncertainty principle are observable only on a submicroscopic scale, as in the case of an electron, for example, whose momentum and position can never be simultaneously known. A related phenomenon, the tunnel effect, allows electrons to penetrate barriers up to 100 atoms thick which, on classical theories, would be expected to stop them. This property has been exploited in making tunnel diodes for ultra-fast switching purposes.

The quantum mechanics of Schrödinger and Heisen-berg was extended in the form of quantum electro-dynamics by P A M Dirac, whose work in the late 1920s forms the basis of present day research. Among other successes, Dirac predicted the existence of a particle which had the same characteristics as the electron, but a positive charge. This particle, the positron, was discovered in 1932.

Many physicists nowadays believe that the four funda-mental forces of nature owe their origin to quantum inter-actions. A force is thought to result when an 'exchange particle' is emitted by one subatomic particle and absorbed by another. In the case of the electromagnetic force, virtual photons are exchanged, while pi mesons give rise to the strong nuclear force. Gravity is thought to originate from interactions involving gravitons, although these have not yet been detected. There is a growing body of experi-mental evidence to support the existence of the intermediate vector boson, the particle responsible for the weak force.

FIELDS, FORCES AND

NUCLEAR ENERGY

Albert Einstein and Robert Oppenheimer. Oppenheimer worked on the atomic bomb; as a result of his participation in the debate about the social consequences of nuclear power, his government security clearance was lifted in 1954.

FORCES AND PARTICLES

Below left: an intersecting storage ring, for observing almost head-on particle collisions. Below: a Cockcroft-Walton particle accelerator, the first important one, in the 1930

Particle physics is the study of the behaviour of the fundamental components of matter. It is the continuation of the research which uncovered the mechanism of the atom and of the nucleus at the heart of the atom. Present research is concerned with investigating the nature of the individual particles which make up the atom and of the many other related particles, recently discovered in experiments at particle accelerators.

The number of different particles which have been identified is about 200. The list includes 'antiparticles', identical in mass but opposite in many of their properties to their particle equivalents. For example, an electron is pushed away by an electrode which has a negative voltage while an anti-electron (or positron) is pulled towards it. This is because the electron carries a negative electric charge, while the positron carries a positive electric charge.

The particles range in mass from the photon (the particle of light), which conveys energy but itself has no mass, to those such as the particles known by the letters N and Z which have a mass three times that of the proton and neutron (these form the nucleus which accounts for most of the mass of any atom). Particles have many other measurable properties such as electrical charge (positive, negative or neutral) and spin. When they interact with one another it has been found that some 'conservation laws' apply dictating what can emerge from the interaction. For example, the particles which emerge from an interaction must carry the same total electric charge as those whic went in, they must emerge with the same mass energy, an so on.

Big differences have been found in the types of particl interaction and they have led to the description of particl behaviour as being controlled by four forces: strong electromagnetic, weak and gravitational. They are easil distinguished because their relative strengths are ver different. The electromagnetic force is a hundred times, th weak force a thousand million times and the gravitationa force over a million, million, million, million, millio million times less powerful than the strong force.

Following the categories of force, the particles themselve are grouped into categories. Those that feel the stron force are called hadrons and those that are sensitive to th weak force but not to the strong force are called leptons Those that are electrically charged feel the electromagneti force and all particles that have mass feel the gravitationa force.

These forces are often referred to as 'force-fields' because at every point in space one could imagine a lin whose length shows the strength of the force at that poin and whose direction shows which way a body would mov under the action of the force. These imaginary lines ar called vectors, and a whole region of space filled wit vectors might resemble the blades of grass in a field blowi in one direction—hence the term 'force-field' or 'vecto

Below: the Aldermaston six million volt Van de Graaff high voltage electrostatic generator. The energy beams produced are used to bombard various substances for nuclear study.

Bottom of page: a section of the world's largest ring accelerator ($1\frac{1}{2}$ km), in the Soviet Union. There seems to be no end to the number of smaller and smaller particles to be discovered.

Below: a linear accelerator, designed to accelerate heavy particles. The outer tube is 90x10 feet (27x3m). The TV picture tube, a cathode ray tube, uses the principles of an accelerator.

field'.

Sometimes it is more useful to know the work needed to move a particle from one point to another rather than the force on it in a field. The potential at a point in a gravitational field is the energy (or work) needed to take one kilogramme from that point to 'infinity' (which is in practice anywhere sufficiently far away for the effect of the field to be negligible). The work required to take an object from one point (A) to another (B) is the difference between the potential A and that at B, multiplied by the mass of the object. In an electric field the potential is the force needed to move a unit electric charge (one coulomb) to infinity, and the work done in moving a charged body in an electric field is the difference in potential multiplied by the amount of charge on the body. The electric potential is measured in volts, and it is the difference in potential (the voltage) between two terminals of a battery which causes a current to flow when a wire is connected between them.

Inverse square law

One property which all force fields share is the inverse square law, first described by Sir Isaac Newton. This means that the effect of a field varies with the square of the distance to the source of the field, and that it decreases as the distance increases. All force-fields can be described by the same general formula, which contains a constant (called G in the case of gravity), the magnitudes of the two forces involved (the masses of the two bodies), and the square of

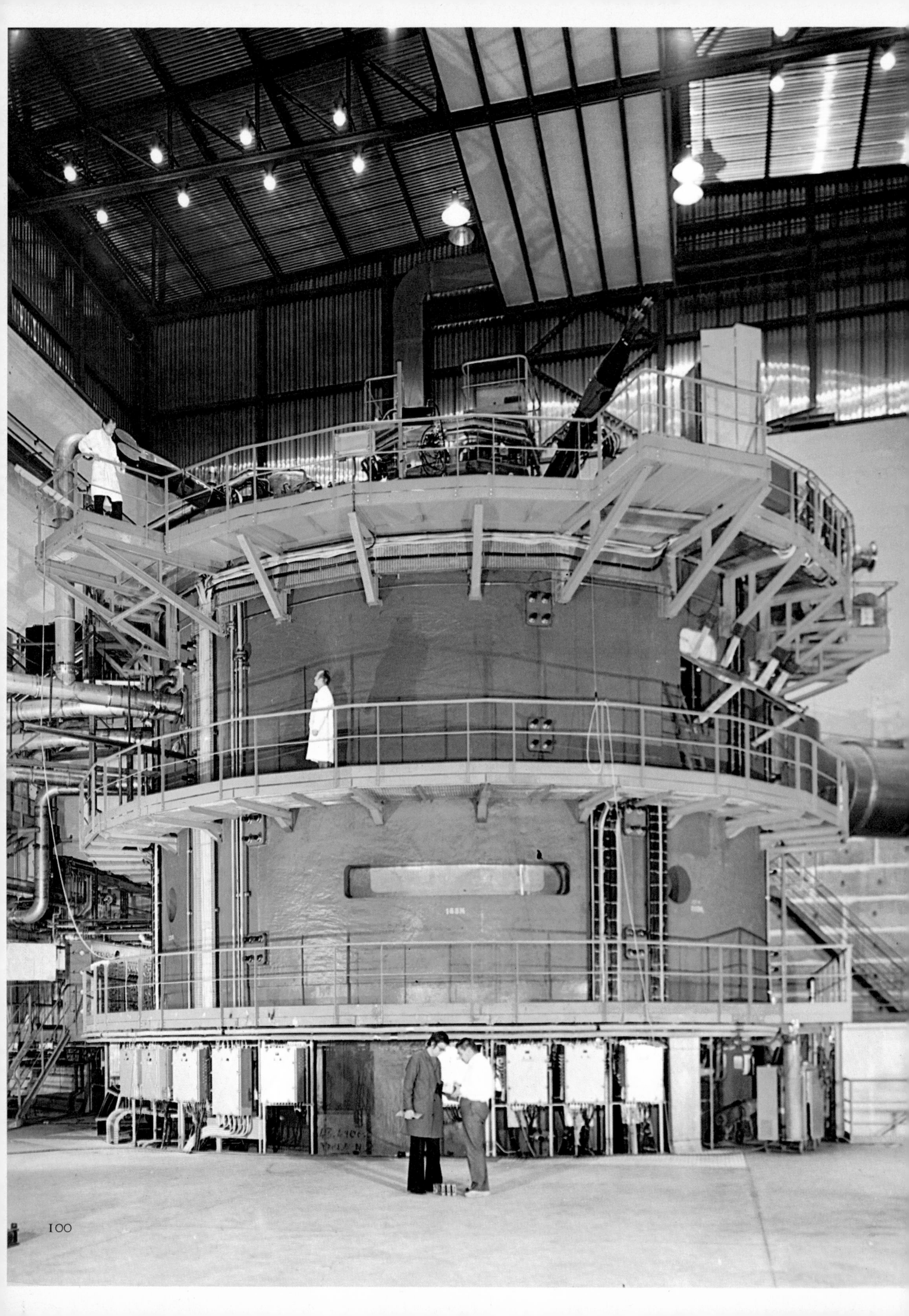

Left: the bubble chamber, an important research tool at CERN, the European nuclear research centre. Using 38 cubic metres of liquid hydrogen, a superconducting magnet, a piston operating from below and four cameras, it makes visible the tracks left by particles of disintegrating atoms. Below: evidence on film of particle behaviour, which can be read by physicists. Top left, for example, was the discovery of the positron in 1932.

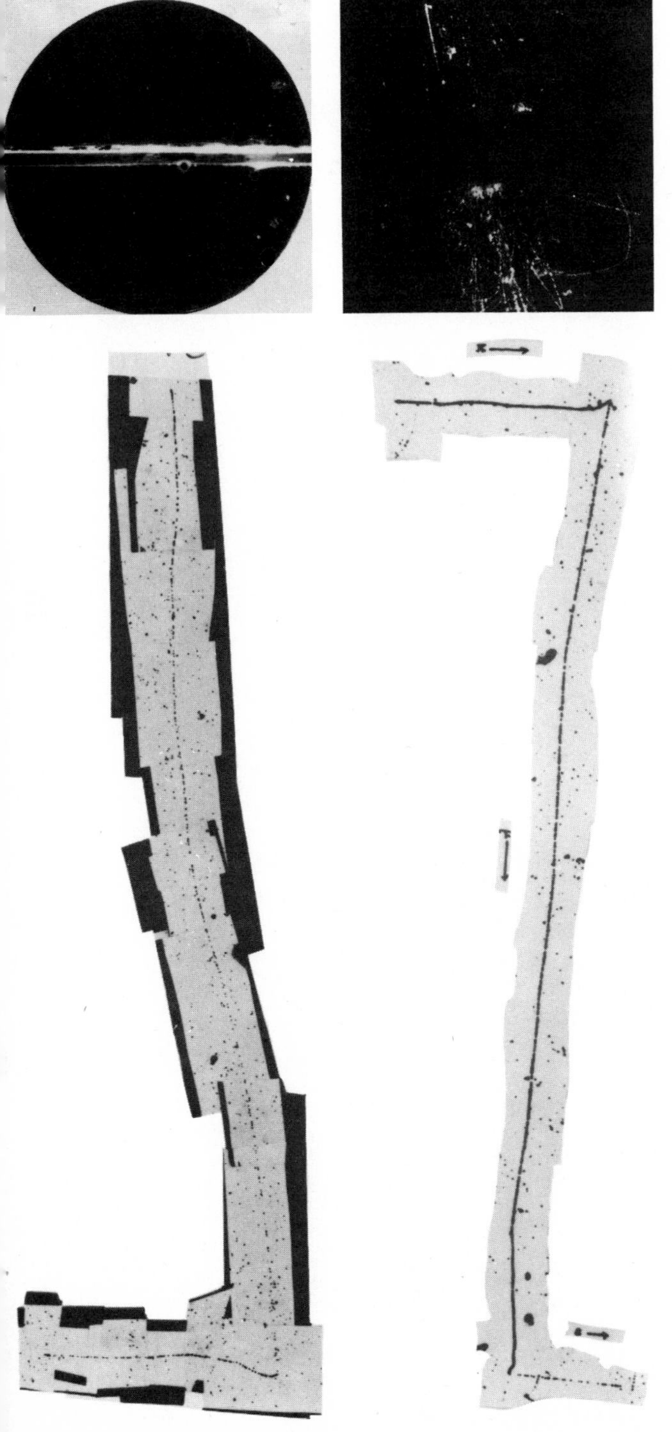

their distance apart.

Gravitational force

Isaac Newton first realized that the planets would naturally move in ellipses if there was an attractive force between the Sun and the planets which depended on the product of the masses of the two bodies divided by the square of the distance between them. He showed that it was the same force which attracts an apple towards the Earth, by comparing the force on the apple with the force needed to keep the Moon in orbit about the Earth. Since the distances from the centre of the Earth to the apple and to the Moon were known, he could demonstrate that these forces also depended on the inverse square of the distance—that is, the force decreases as the square of the distance increases.

Measuring gravity

The force of gravity is actually extremely weak and it is only because the Earth is so massive that its gravitational effects are obvious. The attractive force between two 20 kg (44 lb) objects 30 cm (1 ft) apart is only the same as the weight of one thirty-thousandth of a gramme (one millionth of an ounce) on Earth. The first measurement of gravitational force between two bodies of known mass was made by Henry Cavendish in 1798.

His apparatus consisted of two 2 inch (5 cm) diameter lead balls (each weighing 1.7 lb, 0.75 kg) hung from the ends of a six foot (2 m) long deal beam, which was supported at the centre by a long wire allowing the beam to swing horizontally. Two 12 inch (30 cm) diameter lead balls (each weighing one sixth of a ton) were placed near the small balls on opposite sides, so that the gravitational attraction between each pair of large and small balls caused the beam carrying the latter to swing towards the large balls. The 12 inch balls were then moved to the other side of the small balls, making the beam swing the other way. The total swing measured at the end of the beam was 3/10 inch (8.5 mm), and from this Cavendish calculated the force between the lead balls. He expressed his results as the gravitational force between two 1 kg masses 1 metre apart, a quantity usually called G, Cavendish's value for G was the best for almost a century, and is within one per cent of the best modern value (0.00000000006673 newton, that is 6.673×10^{-11} newton).

Newton's gravitational theory also predicted that all objects at the same place will fall equally fast towards the centre of the Earth. The ancient Greeks, in particular Aristotle, had maintained that heavy bodies always fall faster than lighter ones, but in 1590 Galileo disproved this hypothesis. According to the legend he dropped two objects of different mass from the Leaning Tower of Pisa, and they hit the ground simultaneously. There is actually a very slight difference in acceleration between light and massive bodies, but the Earth is so much larger that this is unnoticeable.

Variations in gravity

The acceleration due to gravity at any place is called 'g', and is about 32 feet per second per second (9.8 m/s²). It changes slightly according to the altitude and latitude of the place where it is measured. The Earth acts gravitationally as

Right: decay of a kaon (1949); annihilation of an antiproton by a nucleus (1958); and at bottom, a bubble chamber photo taken in 1965, with its interpretation next to it.

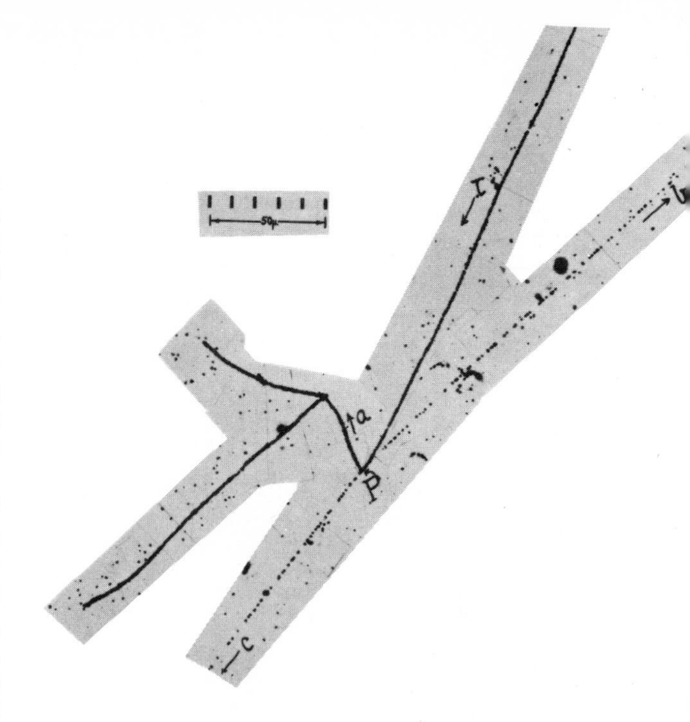

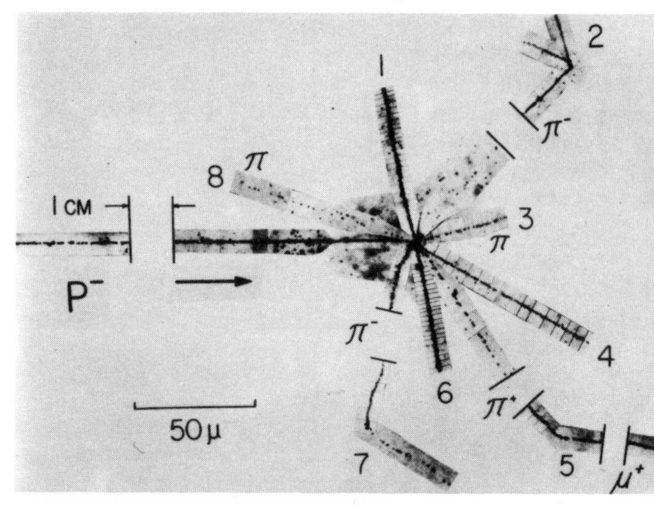

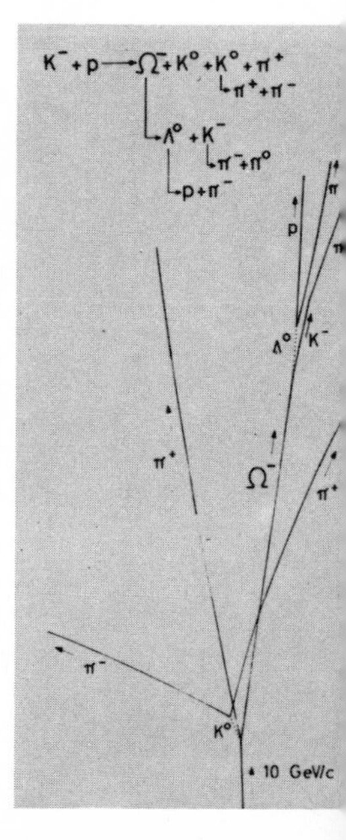

if all its mass were concentrated at the centre. It is not a perfect sphere, so a change in either altitude or latitude means a change in the distance from the centre of the Earth, and thus a change in the gravitational force (according to the inverse square law). At a height of 100,000 feet (30,000 m) 'g' is 99% of its value at sea level. Even at the altitude of an orbiting spacecraft, about 200 miles (320 km), the gravitational force is still only 10% less than at the surface. But the spacecraft is given sufficient orbital velocity that centrifugal force exactly balances that due to gravity, and the astronauts experience weightlessness.

Einstein's General Theory of relativity (1915) introduced a new theory of gravity, which for everyday purposes is the same as Newton's, but it explained a puzzling discrepancy in the motion of the planet Mercury. Einstein's theory also predicted that light as well as matter is affected by gravity, and astronomical observations have proved that this effect does occur. Other predictions of the theory, such as the existence in space of 'black holes' (in which the gravitational field is so strong that light cannot escape) and gravitational radiation (in some ways similar to electromagnetic radiation) are still being investigated by astronomers. Another more recent theory suggests that the value of G may change very slowly over the duration of the Universe, thus accounting for some of its large scale properties such as its apparent expansion.

Electromagnetic force

The electromagnetic force acts between all particles carrying an electric charge. It is seen in action in all phenomena of electricity and magnetism, in light and in radio waves. It controls the motions of the negative electrons around the positive nucleus of the atom and is thus the source of all atomic behaviour. It is attractive between oppositely charged particles, pulling them together, and repulsive between similarly charged particles, pushing them apart. The effect is proportional to the product of the two charges involved and it stretches out from a charge to infinity decreasing as the inverse square of the distance from the charge in a similar way to the gravitational force.

The way in which the electromagnetic force acts is described in a theory known as quantum electrodynamics, QED, which says that each particle is surrounded by a cloud of photons—the packets of energy without mass such as are present in light. The particle is seen as constantly lobbing out these photons and then catching them again. If a photon meets another particle an interaction takes place (such as a pulling together if the particles are oppositely charged); via the photon a particle has passed a message to the other particle. The photon cloud is densest near the particle and thins when moving out from the particle to infinity. This explains how the strength of the electromagnetic interaction falls off with distance.

QED has proved to be a highly successful theory. It can be tested because, if charged particles have a photon cloud, this will slightly modify their properties. These slight changes have been measured and found to agree with the predictions of QED theory to an accuracy of a few parts in a million. The theory was developed to explain a compara-

Below: the inverse square law. Three metres away, radiation has three times the area but is nine times as weak.
Bottom: iron filings on paper around magnets.

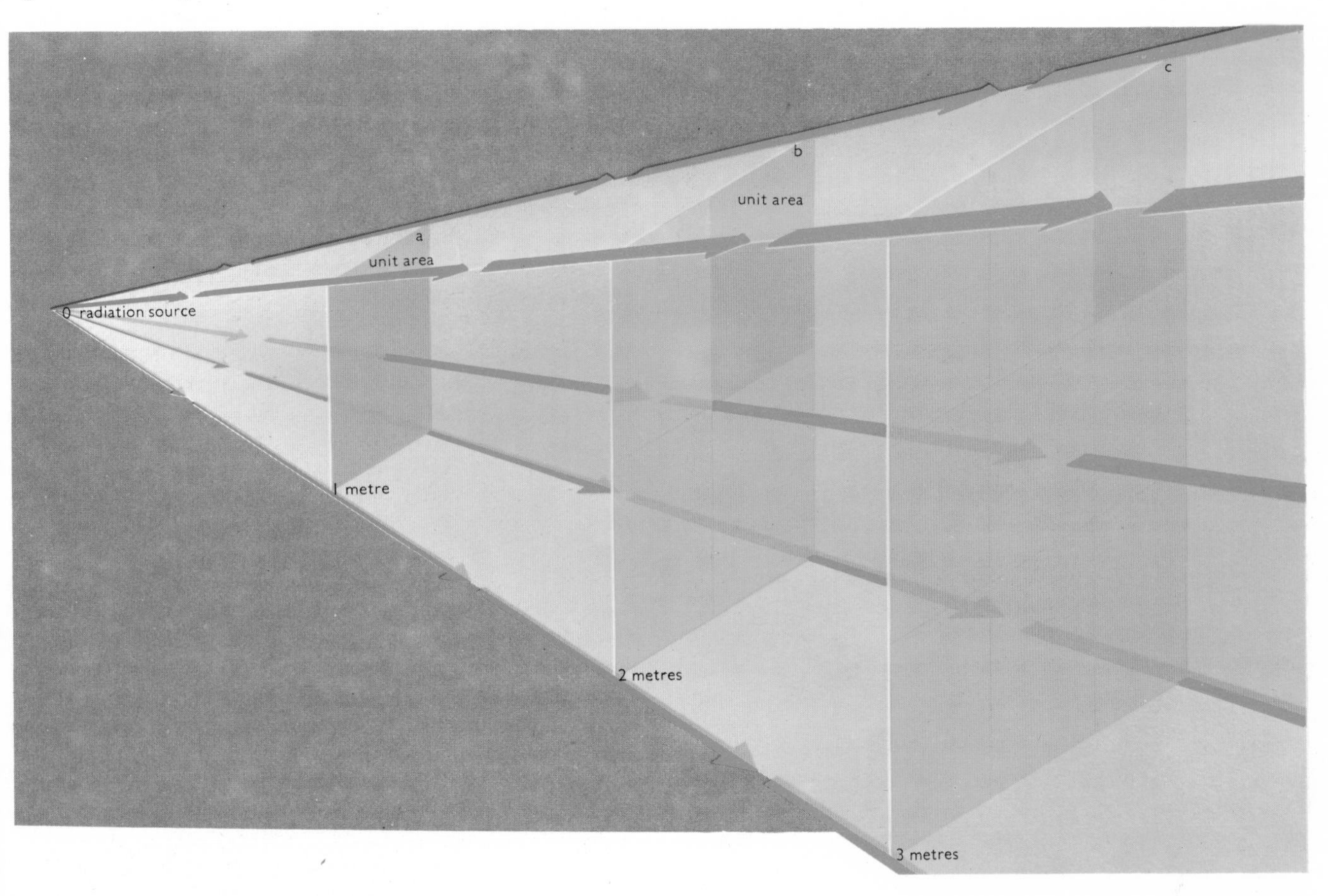

tively small range of phenomena but it has worked perfectly in predicting the effect of the electromagnetic force, whether at distances as small as a thousandth of a millionth of a millionth of a centimetre (deep inside a charged particle) or out to distances as large as a million miles.

Strong force

The strong force acts between hadrons. It can be seen in action in the nucleus of the atom, where it holds clusters of protons and neutrons close together despite the electromagnetic force between the positively charged protons trying to push them apart. Its effect is felt only at short distances of about the dimensions of the nucleus (a millionth of a millionth of a centimetre) which explains why the nucleus does not pull in other hadrons unless they actually enter the nuclear volume.

The way in which the strong force acts is interpreted as being through the exchange of particles known as mesons of which the pi meson (or pion) is the most common example. A hadron communicates its presence to another hadron by exchanging a pion like charged particles exchanging photons. The pion, however, has mass (about a quarter of that of the proton), and the pion cloud sticks very close to the hadron, so the strong force does not spread out far from the particle.

This comparatively simple picture was thrown into confusion by the discovery at accelerators of a multitude of hadrons and mesons. When particles were hit with other

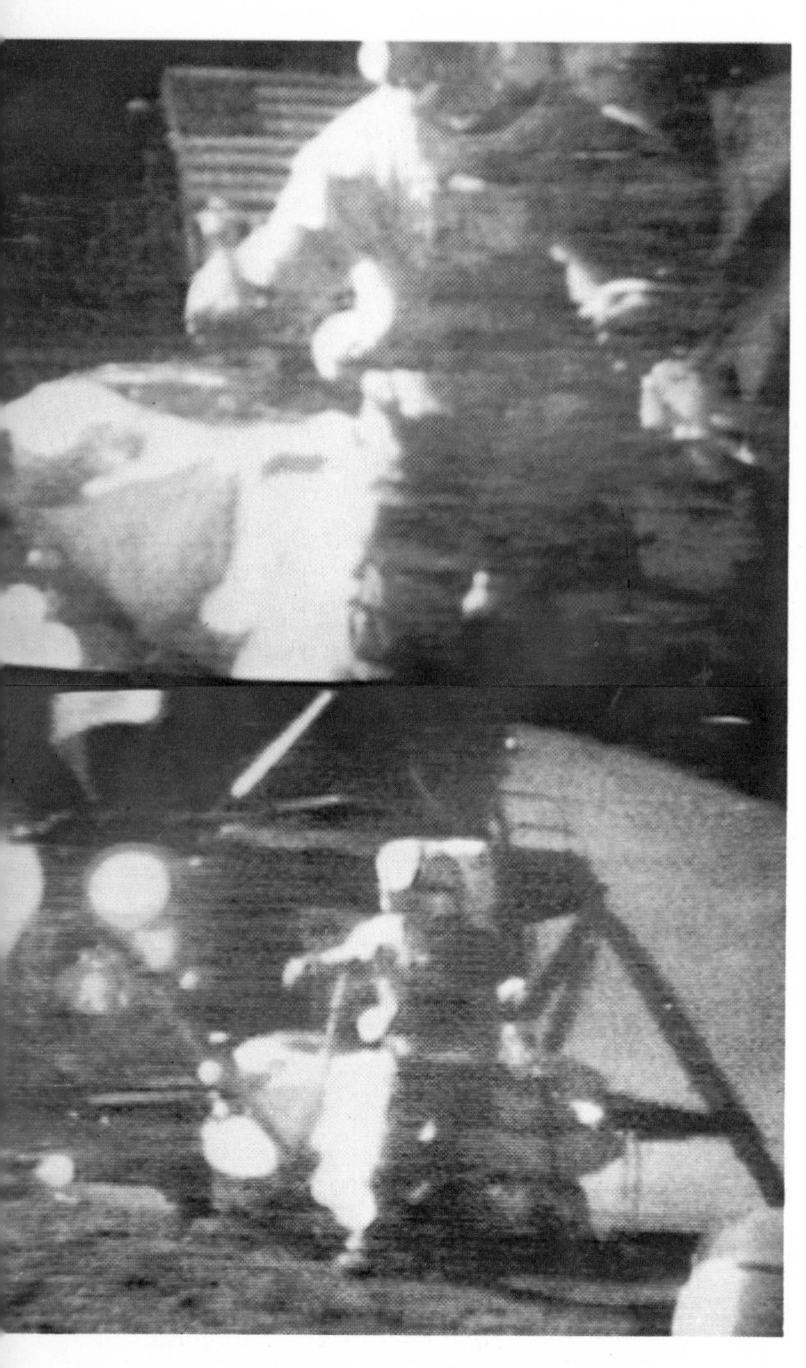

Above: on the Moon, astronaut David Scott dropped a hammer and a feather in a slightly different version of the Tower of Pisa experiment. In the airless atmosphere, they dropped at the same rate.

particles which had been accelerated to high energies many previously unknown particles emerged, all acting under the influence of the strong force: in the early 1960s particle physicists collected new particles rather like botanists collecting new flowers, with no understanding of why they should exist or of what role they play in Nature. As the properties of the particles were determined it was realized that there are orderly relationships between them, rather like those between groups of chemical elements.

Just as the structure of the atom underlies the relationship of the Periodic Table, so it is believed that there is structure in the hadrons leading to their relationships. The hadrons group together in sets of eight and ten particles of related properties and it was deduced that these sets could result from the existence of three more basic objects, given the name quarks, coming together in different ways to build up the hadrons.

Recently, there has been a series of experiments where high energy particles of different kinds have been bounced off protons and neutrons. The aim is similar to the famous experiment of Rutherford at the beginning of this century. He bounced particles off an atom and from the way that they scattered, he was able to deduce the presence of the tiny nucleus at its centre. The scattering of particles on protons and neutrons indicates the presence within these hadrons of three tiny grains—possibly the quarks. It appears, therefore, as if the particles that we had regarded as the fundamental components of matter are themselves composed of more basic objects.

Weak force

The weak force can be seen in action in the break up, or decay, of particles into other particles. The time involved in particle decay can be about a thousand millionth of a second. This is very long on the scale of particle interactions (the strong force acts a million million times faster) and reflects the weakness of the force. As far as is known, the force operates within a particle and does not extend beyond its boundary.

Until the high energy accelerators came into action, most of the knowledge of the weak force came from observation of the radioactive decay of the nucleus—in particular, the so-called beta decay due to the break-up of a neutron into a proton and electron. It became clear that since the electron emerges with a variety of energies some other particle must be carrying away the remainder of the energy from the break-up. This invisible particle was called by physicist Enrico Fermi a *neutrino* ('little neutral one'). It has no charge and is a lepton, feeling only the weak force. W Pauli felt safe, therefore, in betting a case of champagne that such a particle would never be observed. Unfortunately for Pauli, it has been observed interacting with other particles around nuclear reactors, from which millions of neutrinos pour out every second, and at high energy accelerators, where beams of neutrinos are used to study the weak force.

Such experiments have revealed that the behaviour of particles under the influence of the weak force overthrows many 'commonsense' ideas about the laws of Nature. The first overthrow came in the 1950s from measurements on the electrons emerging from the beta decay of the nucleus. They are all observed to be spinning in a clockwise direction, implying that the unseen neutrinos coming out with them are spinning in an anti-clockwise direction. This means that when the weak force is in action Nature is concerned

Above: this apparatus to detect gravity waves monitors the position of a large aluminium cylinder, normally in a vacuum. Claims of detection are unconfirmed.

about directions. Previously it has always been believed that 'right' and 'left' were human conventions, to help us, for example, drive on the safe side of the road, but that Nature did not distinguish between right and left—that Nature did not insist on things happening in a particular direction. Following these experiments on the weak force, it became possible to communicate our 'right' and 'left' to an inhabitant of a remote planet by asking him to observe the spin direction of the electrons coming from beta decay.

Ten years later, observing the weak force in action in the decay of one of a type of meson known as the neutral *kaon* produced a more drastic overthrow of a common-sense idea. It showed that not only was Nature concerned about direction in space, it is also concerned about direction in time. The belief that going backwards in time would see events exactly reversed, like a film running backwards, is no longer valid.

No satisfactory theory exists to explain all the manifestations of the weak force which have been observed. One theory has taken over the idea of the exchange of a particle in a similar way to the photon in the electromagnetic and the pion in the strong force. It is called the intermediate *boson* or W particle but it has never been discovered. The weak force is the most mysterious of those in action in particle physics.

Unified field theory

There are many questions about the nature of fields which science leaves unanswered. Einstein in particular spent much of his life after his work on relativity in trying to find some way to link all these forces, by showing how they might all be due to one basic cause, or might consist of combinations of similar components in the same way that the wide variety of chemical elements can be explained by combining different numbers of neutrons, protons and electrons in atoms. But Einstein's and all other subsequent attempts have failed.

Electromagnetic fields, such as light, take a certain time to have an effect, owing to the speed of light. They can also be broken down into the action of small 'packets' of energy, called quanta. But physicists cannot agree on whether the other fields have similar properties, or, for example, whether anti-gravity might exist in the same way that there are two electric or magnetic poles. Some scientists have suggested further forces, to account for large scale movements in the universe.

MASS-ENERGY EQUIVALENCE

Above: this power station has coal stocks of 140,000 tons. If it could be completely converted into energy, it would make 4000 million million MWh—enough to supply the current energy demands for 30,000 years.

One of the most far-reaching results of Einstein's 1905 Special Theory of relativity is the one which states that mass and energy are equivalent. Mathematically, this is the famous statement $E = mc^2$: the total energy contained in a piece of matter is given by its mass multiplied by the square of the speed of light.

Einstein arrived at this conclusion from the basic assumption that the velocity of light is the same for any observer, whatever his velocity relative to the source of light. This goes against everyday experience, yet it is supported by every test that scientists have applied. Einstein's predictions, made from this assumption, are equally unexpected yet have nevertheless been supported by experiment—the most dramatic example being that of the atomic bomb which uses the fission of a uranium isotope.

Although the equivalence of mass and energy is generally accepted, the problems of actually extracting the energy from matter are immense. The Sun, or a hydrogen bomb, uses fusion of hydrogen atoms to make helium atoms, the difference between four hydrogen atoms and one helium atom being released as energy. This process is much more efficient than that of fission, yet if the difference between the two masses is compared with the total mass, it turns out that in the fusion process only 0.7% of the available energy is released.

Astrophysicists and nuclear scientists are equally interested in the problem of extracting as much energy as possible, and have suggested further processes which may work. By pouring matter down a *black hole*, that is a region in space where the matter is so condensed that nothing, not even light, can escape from it, it may be possible to extract as much as 43% of the rest mass of a particle as energy—the particle gives out radiation as it accelerates. No black holes are so far definitely known to exist, though there is no theoretical objection to them. But the only way of converting 100% of the rest mass into energy appears to be complete annihilation of matter by allowing it to meet antimatter.

Small amounts of antimatter—positrons, or anti-electrons— have been created, and when these interact with ordinary matter, all their mass indeed appears as energy. But the anti-matter had to be created in the first place, causing an original loss of the same amount of energy. No reserves of antimatter are known in the universe, and unless some are found, it appears that the dream of completely converting a small amount of matter into enough energy to meet all of mankind's needs must remain an impossible one.

BINDING ENERGY

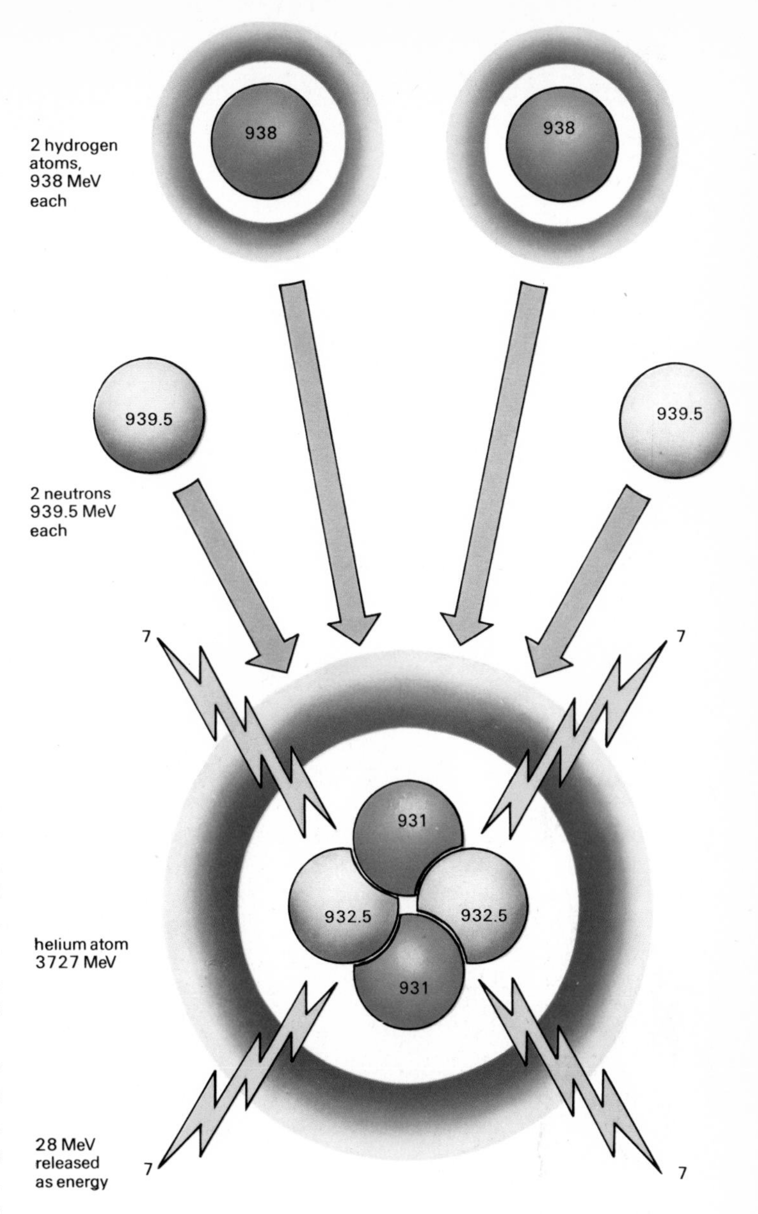

Above: the fusion of two hydrogen atoms to make a helium atom. The two hydrogen nuclei, each consisting of one proton, join with two neutrons to make the new nucleus.

Binding energy is the energy involved in holding together the component parts of a molecule, an atom or a nucleus. It is a term used particularly in connection with the nucleus and is the source of both types of nuclear energy, fission and fusion, which have had such an impact on the modern world.

The realization that large amounts of energy are available from the nuclei of the atoms followed from precise measurements of nuclear masses early this century. In instruments called mass spectrographs, the masses can be accurately calculated by measuring the curvature of the paths taken by charged clusters of particles as they travel through electric and magnetic fields. The measurements reveal something which, taken by itself, seems impossible. The masses of the nuclei are less than the total arrived at by adding the masses of the protons and neutrons of which they are formed.

This enigma can be resolved only through Einstein's famous formula $E = mc^2$ (energy equals mass multiplied by the speed of light, squared). Mass and energy are interchangeable and the mass which is missing in the sums on the nucleus is related to the energy which holds the nucleus together. This is the binding energy, which was also known initially as the 'mass defect'. Some of the particles' mass has been converted into energy and to separate the original particles, with their higher total mass, energy has to be supplied to the nucleus to break it up.

When the nuclei of all the chemical elements are examined a crucial fact emerges. Most of the elements have nuclei with a binding energy of about 8 MeV or 8 million electron volts per particle—nearly a thousand times the energy of an electron 'fired' by a television picture tube. The lightest and heaviest elements, however, have less binding energy per particle on account of the structure of their nuclei. The amount of mass that is 'missing' when doing the sums on the nucleus is greater for the elements near the centre of the periodic table; if the light or heavy elements can be converted into these central elements some mass can be liberated as free energy.

To take a specific example, a helium nucleus is built up of two protons and two neutrons. The mass of a proton (the nucleus of the normal hydrogen atom) is about 1.7 millionth of a millionth of a millionth of a millionth of a gramme. This is a rather clumsy unit with which to calculate and, using the mass-energy equivalence, the nuclear physicist usually handles the proton mass in units of energy—a proton is 938 MeV, and a neutron is slightly heavier with 939.5 MeV. The helium nucleus is, however, 3727 MeV or 28 MeV less massive than its constituents: 28 MeVs worth of energy can be released in the formation of a single nucleus. Compared with conventional sources this is a colossal amount of energy.

Conventional energy sources, say the burning of oil, involve atomic binding energies—they are chemical reactions to do with the binding of the electron clouds surrounding the nuclei in atoms and molecules. Here also, the release of energy is connected with the conversion of mass—the products of the burned oil have very slightly less mass than the oil itself—but nuclear energy sources, which bring the centre of the atom into play, are several million times more powerful. If all the hydrogen nuclei in a glass of water could be combined to create helium, enough energy would be liberated to drive an ocean liner across the Atlantic.

This process of joining light elements together to form heavier ones is known as fusion, and it is the process that powers the hydrogen (thermonuclear) bomb and the stars. There is a great deal of research under way in an attempt to master fusion in the laboratory so that it can be used in power plants.

At the other end of the table of elements, the breaking up of heavy nuclei such as uranium is another source of energy. This process is known as fission. A uranium nucleus usually breaks up into two almost equal parts (which end up, for example, as nuclei of the elements molybdenum and palladium). Because of the different binding energies of heavy and light atoms about 200 MeV is liberated in such a fission, and this powers the nuclear reactor and the nuclear bomb.

RADIOACTIVITY

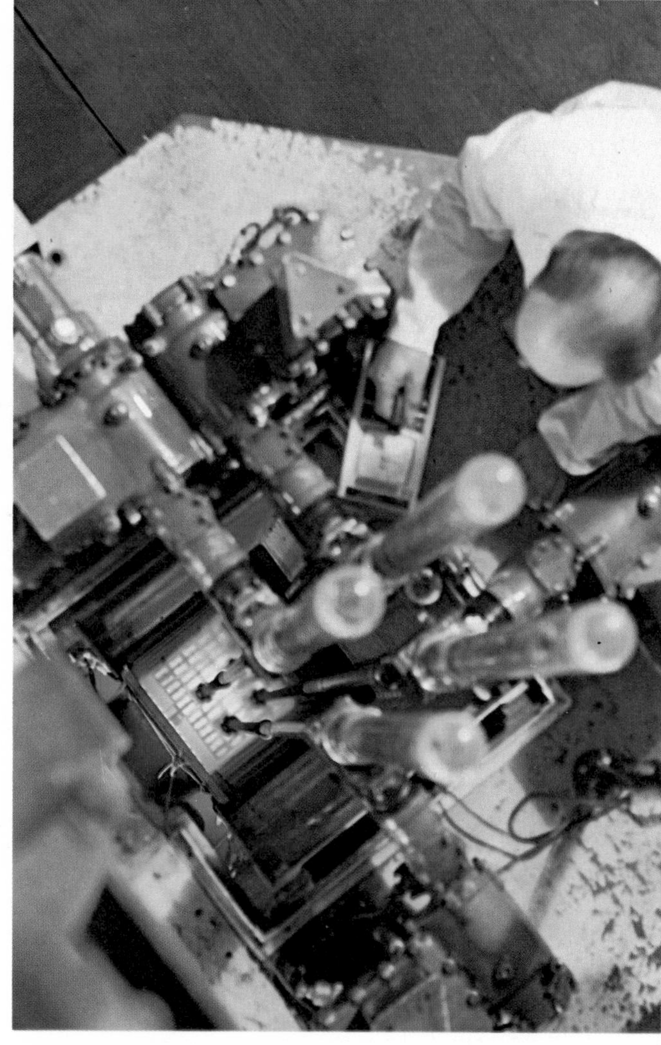

Radioactivity is a property of certain kinds of chemical elements whose atomic nuclei are unstable: in time each such nucleus achieves stability by a process of internal change called radioactive decay, which involves a release of energy in a form loosely known as 'radiation'. The energy involved is very large by comparison with that released by chemical reactions involving the same amount of material, and the mechanism by which it is released is totally different.

Radioactivity was discovered in 1897 by the French chemist Becquerel, during his studies of fluorescence. He found that an unexposed photographic plate wrapped in black paper was affected as if by visible or ultra-violet light (or by Röntgen's newly discovered X-rays) when the package was placed in contact with compounds of the heavy element uranium. He deduced (correctly) that some form of radiation must be coming from the uranium and penetrating the paper to reach and affect the photographic emulsion. Careful study by Becquerel and other scientists, including the Curies, Joliot, Soddy, Rutherford, Chadwick and Geiger, revealed that a number of heavy chemical elements, many of them previously undetected because of their rarity, appeared to be internally unstable and gave off penetrating radiations. In the process, they themselves changed into different elements following intricate but well-defined paths to eventual stability. This phenomenon, entirely unlike anything previously encountered, was given the name radioactivity, and the process of change was called radioactive decay.

The nuclear atom

In the atomic nucleus there are two sorts of particle, protons and neutrons. The protons each carry a single positive charge and, because they can bind to the atom an equal number of negatively charged orbiting electrons, their number governs all the atom's chemical properties. This number is called the atomic number and is symbolized by the letter 'Z'. Z ranges from one for hydrogen (including its isotopes deuterium and tritium) to 92 for uranium, and for identification purposes can be regarded as an alternative to the element's name. The neutrons are electrically neutral and therefore have no effect at all on the chemical properties of the atom. The number of neutrons ranges from zero in ordinary hydrogen (and one and two respectively in its two isotopes) to 146 in the heaviest natural (uranium) isotope.

The sum total of protons and neutrons in a nucleus is called its mass number (designated 'M') and, written after the element name, is used to identify a particular isotope: carbon-14, for example, is the carbon isotope with 6 protons and 8 neutrons. Only certain combinations of Z and M give nuclei that are stable: if there are too many or too few neutrons the nucleus will sooner or later undergo a change, radioactive decay, which will, in one or more steps, bring it to stability. The degree of instability shows up in the energy emitted in the decay process, and also in the decay rate. The latter is measured in terms of half life or halving time, which is the time taken for half the number of atoms initially present to have undergone the decay process. Halving times range from fractions of a second to millions of years. There are several different ways in which radioactive decay can take place: alpha decay, beta decay and gamma ray emission.

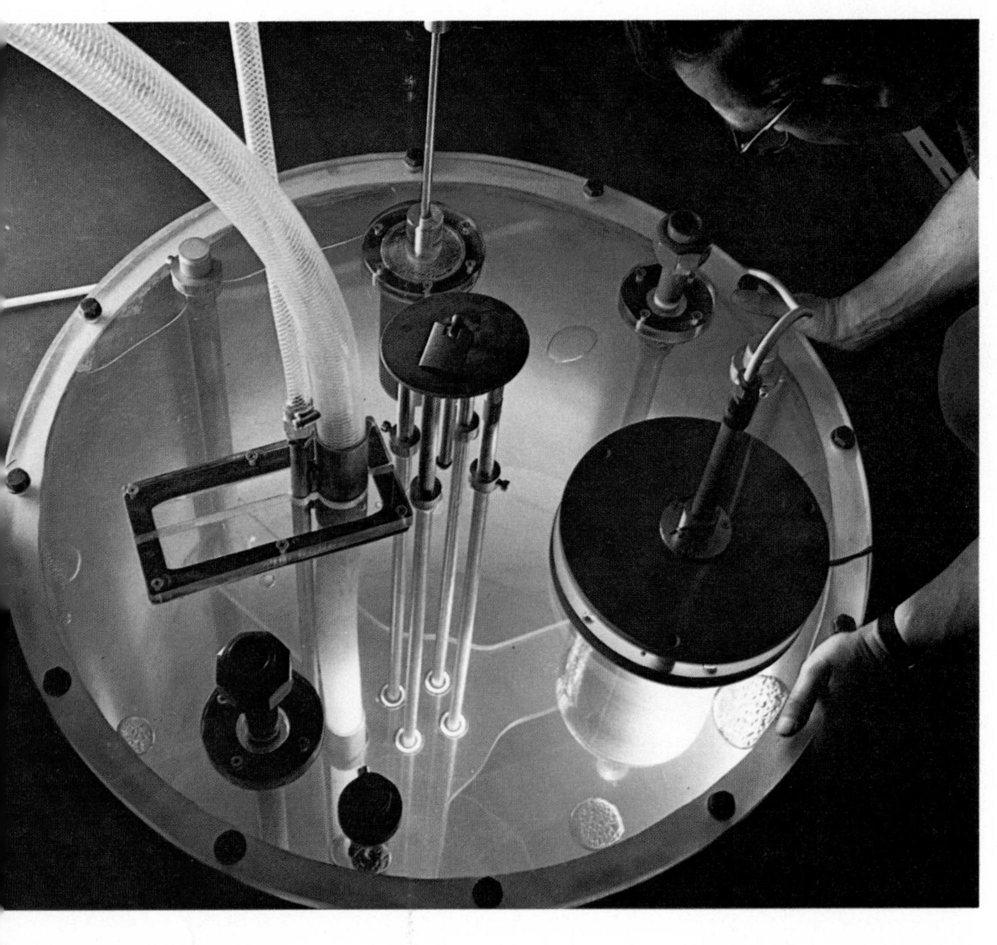

Far left: inserting a fuel pin into an element for an advanced gas-cooled reactor. The 36 fuel pins each contain pelleted uranium oxide powder surrounded by a graphite sleeve.
Centre: the light produced by high energy beta particles (electrons) entering a transparent medium was discovered by Cerenkov in 1934 and can be used to detect particles. *(See also pictures on pages 51 and 52.)*
Left: a californium-252 installation for activation analysis.

Alpha decay

n a nucleus that is too heavy to be stable a compact group an alpha particle) consisting of two protons and two neutrons is ejected, leaving the nucleus four units lighter n M and two units lower in Z, in other words two steps lown in the periodic table. Structurally, an alpha particle s identical to a helium-4 nucleus. Alpha decay is common among the heaviest natural elements (uranium, polonium and radium for example) but does not lead directly to stable nuclei. Intermediate isotopes are produced first.

Alpha particles have energies of up to five million electron volts (a common unit of energy when dealing with particles, abbreviated eV), but they are so bulky that they only pass hrough an inch or so of air and can be stopped by a sheet of paper or the outer layers of the human skin. For that very reason, however, they can cause serious internal damage when emitted by alpha-active materials that have been inadvertently absorbed by the body as airborne dust or through contaminated wounds. The natural alpha emitters such as radium are of limited practical use now that a wide variety of artificial radioisotopes are freely available. However, uranium and its artificial by-product, plutonium (another alpha emitter), are also both fissile and of supreme importance in nuclear power production.

Beta decay

In a nucleus with too many neutrons, one neutron changes to a proton plus an electron, the latter being ejected from the nucleus. An electron emitted in this way is called a beta particle. The nucleus is left with an additional positive charge, and is therefore one unit higher in Z and one step up in the periodic table. Beta particles have maximum energies ranging from 0.02 MeV (million electron volts) to 5.3 MeV, and penetrate up to several metres of air, a few centimetres of tissue or several millimetres of metal or plastic (which provides adequate shielding). They can cause severe surface burns or serious internal harm, especially if emitted inside the body for long periods. Beta decay is the commonest mode of radioactive decay, both among artificial isotopes and among the radioactive products of natural alpha decay. A few of the artificial radioisotopes made in particle accelerators, or separated from the fission products formed in nuclear reactors, have too few neutrons rather than too many. These decay by emitting positrons (positively charged electrons) which interact almost immediately with ordinary electrons to produce 'annihilation radiation' of 0.51 MeV energy, with the qualities of gamma rays (see below). Isotopes which emit positrons (for example technetium-147) have applications in medical diagnosis.

Gamma ray emission

This occurs whenever beta decay has not carried away enough energy to give complete stability to the nucleus.

Above: pitchblende, in which the Curies discovered radium. It also contains uranium, but was too radioactive for that.

Many natural and artificial alpha and beta active isotopes are gamma ray emitters. Gamma rays are a form of electromagnetic radiation like X-rays, having energies between 0.15 and 2.5 MeV. They are reduced in intensity by passing through matter to an extent that depends on their own energy and on the physical density of the absorbing matter. Gamma rays are not stopped in the way that alpha or beta particles are, nor are any materials opaque to them as to light. From 5 to 25 cm (2 to 10 inches) of lead, or up to three metres (yards) of concrete, may be needed to provide adequate shielding against high energy sources of gamma radiation. Excessive external gamma radiation can cause serious internal damage to the body, but cannot induce radioactivity in it or anything else.

Other modes of radioactive decay include internal conversion, where a reorganization within the nucleus results in the emission of X-rays; and electron capture, where a nucleus with too many protons captures an electron from an inner orbit of the same atom, converting a proton to a neutron, with the emission of X-rays and a drop of one place in the periodic table. The nuclei of the alpha emitters uranium-235 and uranium-238 very occasionally decay by spontaneous nuclear fission, producing any pair of a range of possible fission product nuclei, and free neutrons. The artificial radioisotope californium-252 decays exclusively by spontaneous fission and provides a useful source of neutrons. A few fission product isotopes, notably iodine-122, decay by delayed neutron emission soon after they are formed and play an important role in the control of reactors.

The modes of decay, the halving times and the energies of emission (maximum energies in the case of alpha and beta particles) are together uniquely characteristic of the particular isotopes involved: they can be used in the identification and measurement of the emitters themselves, and hence of their precursors, by the technique known as activation analysis.

FISSION AND FUSION

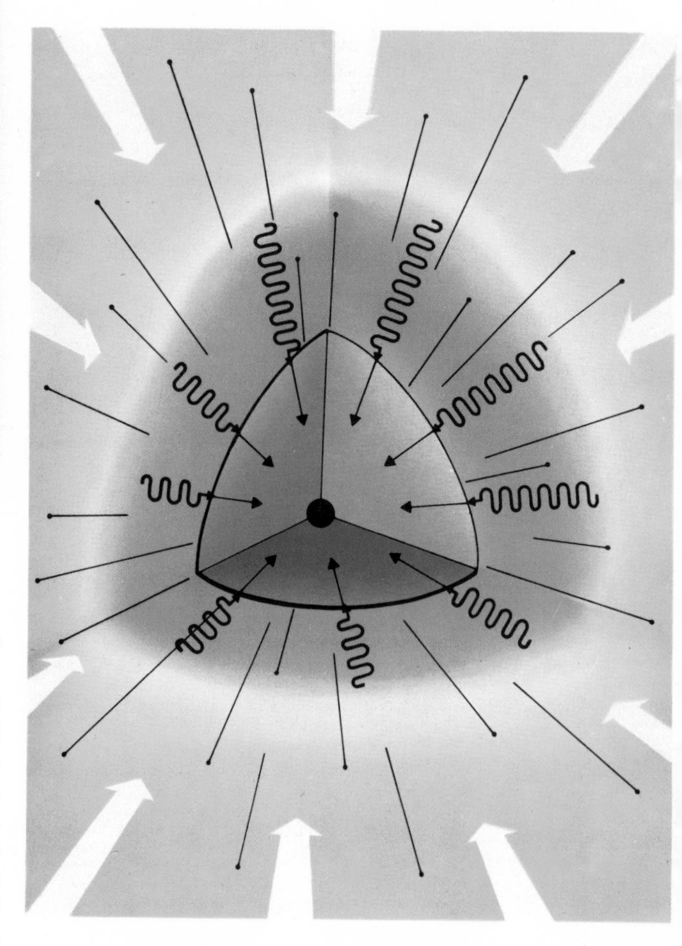

Nuclear fission is the breakup of the nucleus of a heavy atom into lighter atoms. Because of the binding energy which is required to hold the particles together in the heavy nucleus, the fission of such a nucleus releases energy.

Just before World War II several groups of scientists were studying the behaviour of uranium, the heaviest of the elements found in Nature. The uranium nucleus contains 92 positive charged protons and a varying number (about 145) of neutral particles called neutrons. They looked at what happend when the uranium nucleus absorbed an additional neutron, and repeatedly found traces of much lighter nuclei such as those of barium. It was O Frisch and L Meitner in 1939 who first put forward the idea that the uranium nucleus was so disturbed by the additional neutron that it broke or fissioned into two approximately equal halves, resulting in other much lighter nuclei. Doing the sums on the masses of the initial and the final nuclei reveals that some mass disappears when fission takes place. It reappears as energy (since mass and energy are interchangeable) and the lighter nuclei or fission fragments tend to fly off with quite high velocities. The energy liberated in the fission of a uranium nucleus is fifty million times greater than that liberated when a carbon atom unites with oxygen atoms in the burning of coal.

The process has been described as similar to the behaviour of a large drop of liquid to which a tiny bit more

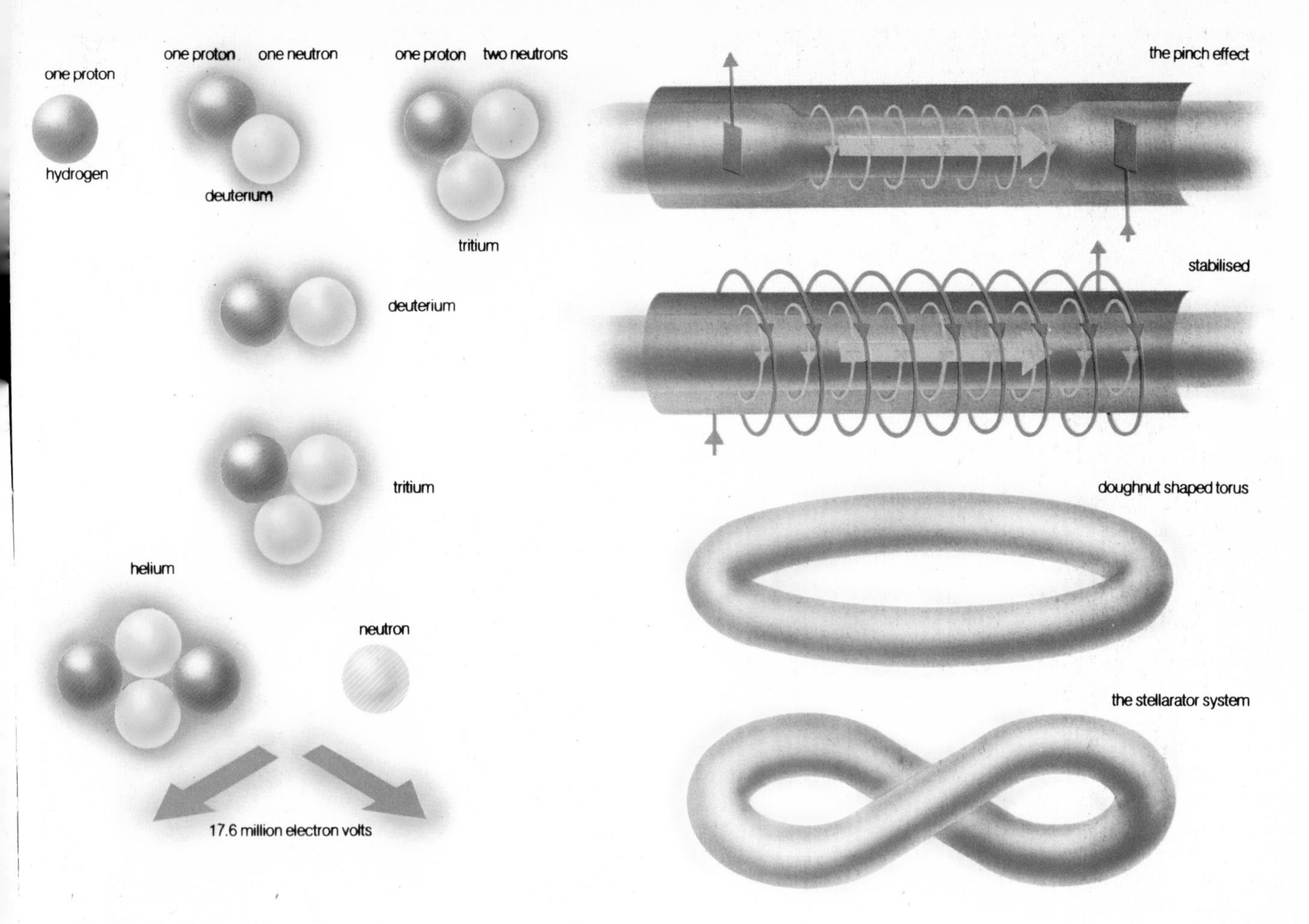

Left: nuclear fusion by laser implosion. Light is absorbed at the critical surface and energized matter travels outwards, but the reaction from this force acts inwards, imploding the remainder of the pellet.
Above: nuclei of deuterium and tritium combine to make one of helium, releasing a free neutron. Diagrams show methods of magnetic containment of gas plasma in which fusion takes place; shapes are common for plasma containment.

liquid is added. The forces holding the surface of the spherical drop together are no longer able to retain the volume of liquid; the drop distorts into dumb-bell shape and splits into two smaller drops.

A crucial feature of the fission process is that the heavy nucleus is so loaded with neutrons that a few neutrons can break off separately as the fission takes place and the lighter nuclei can be too 'neutron-rich' and tend to shed further neutrons in settling down to a stable state. This liberation of neutrons in the fission process opens the door to a chain reaction because they can be absorbed by other uranium nuclei, initiating fission, liberating further neutrons and so on.

An uncontrolled chain reaction is the source of the colossal energies of the atomic bombs. The fission of heavy nuclei is allowed to propagate extremely quickly and each breakup liberates energy. A controlled chain reaction is the source of energy in the nuclear reactor. Here the number of neutrons available to initiate further fissions is kept under precise control by using materials such as cadmium which readily absorb neutrons.

The uranium 235 nucleus, with 92 protons and 143 neutrons, is the only one naturally occurring in Nature which can be induced to fission by sending in a low energy or 'slow' neutron. It was the first choice for use in nuclear reactors. Its much more abundant relative, uranium 238 with 147 neutrons, needs a fast neutron to disturb it enough for fission to occur. Other heavy nuclei susceptible to slow neutrons are uranium 233 and plutonium 239. These are artificially produced from thorium 232 and uranium 238 nuclei, respectively, when they are bombarded with neutrons.

Uranium 238 can exhibit spontaneous fission. It can break up of its own accord without needing a jog from an incoming neutron. This occurs, however, a million times less frequently than its 'favourite' way, that is, settling to a more stable state by throwing out a small cluster of two protons and two neutrons. Another nucleus exhibiting spontaneous fission is plutonium 240. This nucleus is produced in the hotbed of nuclear transformations which go on in reactors and account has to be taken of its inclination to spontaneously fission in order to ensure reactor safety.

Fusion

Nuclear fusion is the joining together of two nuclei of light atoms to form the nucleus of a heavier atom. Because of the difference in binding energies required to hold together the particles in the different nuclei, the process can release energy.

The most important fusion reactions involve the welding together of the lightest of all the nuclei—those of the hydrogen atom—to make a helium nucleus. In its normal form, the hydrogen nucleus contains a single particle, the proton, but there are variants, or isotopes, where the proton is joined by one or two neutrons. These variants are called respectively the deuteron and the triton.

In terms of energy release the most productive fusion is that of deuteron and triton. It produces a helium nucleus, a

Right: a neon plasma fusion experiment. High temperatures cause the neon to become incandescent.

Below right: apparatus for fusion research. A stream of energized hydrogen atoms help to heat the plasma.

neutron and 17.6 million electron volts of energy. The energy liberated can be calculated by doing the sums on the masses of the nuclei; the helium and neutron have less mass than the deuteron and triton, which because mass and energy are interchangeable has been converted into energy. Compared with normal chemical reactions, where only tiny amounts of mass are converted, the quantities of energy released are enormous. If the hydrogen nuclei in a gallon of water could be combined in fusion reactions they would yield more energy than the burning of a million gallons of oil.

Since the supply of hydrogen is almost limitless (in combination with oxygen in water), fusion offers the prospect of a limitless source of energy. Fusion reactions, however, are difficult to bring about. The proton in each hydrogen nucleus carries a positive electric charge and, just as like poles of magnets repel one another, so these positive charges repel one another and the repulsion becomes stronger the closer they come together. To achieve fusion, they have to unite in a tiny volume about a millionth of a millionth of a centimetre across and they must fly at one another very fast to get so close together, overcoming the repulsive effect of their electric charges. Two factors influence whether productive fusion reactions occur—the closeness with which the nuclei are packed together (the density of the hydrogen) and the velocity with which they are moving (the temperature). Typically temperatures of many millions of degrees Centigrade are needed.

This is the problem—to establish a dense volume of hydrogen at a temperature of millions of degrees. Research centres in many countries are working to achieve controlled fusion mainly by trying various methods to contain a dense hot 'plasma'—a volume of hydrogen consisting of nuclei and electrons which are torn off the atoms at high temperatures. Magnetic fields compress the plasma to produce the necessary densities and to keep the plasma away from the walls of a containing vessel, which would otherwise melt. Another technique is to fire high power laser beams at pellets of solid deuterium. However, a considerable amount of work is still to be done before fusion can be used in power plants.

Fusion reactions have been created in short intense bursts in hydrogen bombs using other explosives to create the necessary temperature and density conditions. Energies equivalent to many millions of tons of TNT are produced.

Despite the difficulty of bringing about man-made fusion, it is, in fact, a very common process in Nature at large. The light and warmth of day are the result of fusion because it is the energy mechanism of the sun and all the other stars.

In this case it is the gravitational effect of large masses which pulls the hydrogen nuclei together. At the centre of our sun, for example, the density of nuclei is about 150 grammes per cubic centimetre and the temperature is about 14 million °C. Even then only one in ten thousand, million, million, million collisions between nuclei results in a fusion. At this slow rate, fusion is converting the hydrogen of the sun into helium, creating the conditions which make life on earth possible.

Henri Becquerel

Henri Becquerel.

Henri Becquerel is famous for his discovery of radioactivity—the radiations given off spontaneously by certain atoms with unstable nuclei. He was born into a family with a tradition for original scientific research and it was while continuing research begun by his father that he stumbled upon this new and exciting phenomenon.

Becquerel's grandfather, Antoine César (1788–1878), had been a pioneer in electrochemistry and had experimented with telegraphy and magnetism, while his father, Alexandre Edmond (1820–1891), had studied light and phosphoresence. Born in Paris, Becquerel studied at the Ecole Polytechnique, entering the Ecole des Ponts et Chaussées in 1874. In 1878 he became an assistant at the Musée d'Histoire Naturelle, often taking his father's place.

For his doctorate, bestowed in 1888, he submitted a thesis on the absorption of light by crystals, which he investigated while carrying on the work begun by his father into phosphorescence and fluorescence. (Phosphorescence is a glow emitted by a substance, which unlike fluorescence continues long after the source of excitation, for example sunlight, is removed.)

By the late 1880s academic accolades were beginning to fall upon Becquerel. In 1889, he was elected a member of the Académie des Sciences, and in 1892 he became Professor of Physics at the Musée d'Histoire Naturelle, as his father and grandfather had been. From 1895 he was a professor at the Ecole Polytechnique and at the same time chief engineer in the bridges and roads department.

The work for which Becquerel is best remembered, the discovery of radioactivity, was brought about by Wilhelm Konrad von Röntgen's 1895 discovery of X-rays, which he first noticed because of their ability to make certain materials fluoresce.

Fascinated by this discovery, Becquerel wondered if the natural luminescence, or glow, of certain minerals might be accompanied by similar X-ray emission. He took a thin, transparent crust of potassium uranyl sulphate (a uranium salt) and placed it on a photographic plate wrapped in lightproof paper. Both plate and crystals were exposed to sunlight for several hours. When developed, an outline of the crystals appeared on the photographic plate. To exclude the possibility of a chemical reaction caused by vapours, a variation of the experiment was tried using a thin sheet of glass to separate the uranium salt from the paper-covered plate, and again the outline appeared. Subsequently, Becquerel found that in a similar experiment where some uranium crystals had been left, unexposed to light, in a dark drawer together with a photographic plate, they left an even stronger outline on the developed plate. So, accidentally, it was discovered that the photographic effect of these crystals was not tied up with any fluorescence arising from exposure to sunlight or X-rays. Further research proved that the uranium in the salt was the active agent.

During the following years, Becquerel became engrossed in his work on this strange radiation. He discovered that it was not, like visible light, subject to the laws of reflection, and he also found that it could discharge an electroscope, a device for storing static electricity. Together with Ernest Rutherford he realized that more than one type of radiation was involved, and in 1900 his experiments on the influence of electric and magnetic fields had proved that one component consisted of beta particles, or high speed electrons, which because of their velocity, have a certain power of penetration.

Becquerel failed to identify the alpha particles, or helium nuclei, in the radiation, but in 1903 he confirmed that a third component was particularly penetrative, being able to pass through several inches of lead. Now known as gamma rays, they were formerly called Becquerel rays, although the original discovery was made in 1900 by P. Villard. Becquerel's enthusiasm for his research into radioactivity influenced his friends the Curies, and they discovered new radioactive elements, more powerful than uranium. In 1901, together with Pierre and Marie Curie, Becquerel suggested that radioactivity might be caused by changes occurring within an atom.

For their researches into radioactivity, these three scientists were jointly awarded the Nobel Prize for Physics in 1903. In appreciation of his important findings, Becquerel was made a foreign member of scientific societies in many different countries, and a Fellow of the Royal Society, London, shortly before his death at Croisac in Brittany.

Albert Einstein

Albert Einstein, the theoretical physicist famous for his theory of relativity, was born in 1879 at Ulm, Germany, to parents of German-Jewish descent. His early education, at first in Munich and later in Switzerland, did not mark him out as any kind of genius at all. He found the pedantic methods of the German education system totally uninteresting, and he used often to miss lessons in order to indulge his voracious appetite for reading. At 17, he entered the Swiss Federal Polytechnic in Zürich after rapid cramming of a subject he had somewhat neglected—mathematics. The more liberal and easygoing atmosphere of Switzerland was much more in keeping with his own reflective ways, and in 1901 after graduating, he took out Swiss citizenship. Years later, in the winter of 1932—3, while Einstein was on a fortuitous visit to the California Institute of Technology, Hitler came to power in Europe, and Einstein decided to stay on in the United States, becoming an American citizen in 1940.

Above: Einstein in 1934; right, an informal seminar.

He had entered the Polytechnic with the idea of becoming a teacher but had difficulty in finding a post and ended up in the patent office in Berne. In 1905 he received his doctorate in physics from the University of Zürich. That year saw the publication of three of his papers, any one of which would have earned him a doctorate. Unfortunately, many of his ideas were only intelligible to a small minority of physicists and recognition did not come immediately.

In each of these papers, however, Einstein had managed to combine the principle

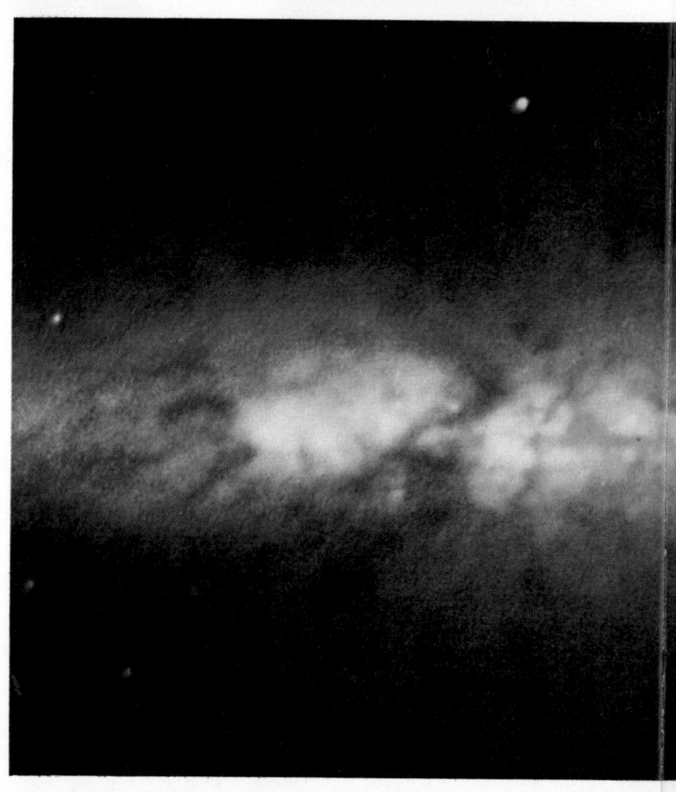

Left: disturbances in space: relativistic motion?

of sound empirical science—that any explanation must make as few unproven assumptions as possible – with an acute insight of new ideas. One of these papers was on the Special Theory of relativity, which was to appear in its final form in 1915 as his General Theory of relativity. It literally overturned the then accepted basis of physics. Experiments measuring the speed of light had shown its speed to be exactly the same whether the source was moving away from or towards the observer. This was, however, a contradiction of Newtonian mechanics.

This paradox was resolved by Einstein's Special Theory. It involved, however, a new interpretation of mechanics, which resulted in the essential equivalence of mass and energy. Matter, said Einstein, was a form of energy, and energy a form of matter, with their inter-relationship expressed by the famous equation $E = mc^2$, where E = energy, m = mass, and c^2 = the square of the speed of light. When a minuscule amount of mass is annihilated, it is replaced by an immense amount of energy: this is the principle behind atomic energy.

The other papers published in 1905 dealt with his explanations of the photoelectric effect, for which he was awarded the Nobel Prize in 1922, and of Brownian motion. The former was a turning point in modern physics. It not only provided the first clear and qualitative account of the ionization of metals by a beam of light, but it also showed that light came in discrete 'packets', which he

Above top: an early picture of Einstein, who was 26 when he published his first paper in 1905.
Above: Henri Poincaré (1854-1912) coined the term relativity, and pointed out that nothing could travel faster than light.

called photons. Light was therefore shown to be 'quantized', just as matter was known to be. Five years previously Niels Bohr had shown a similar quantizing of the energy of the electrons in the atom, and so the stage was now set for the development of the quantum theory. Relativity and the quantum theory are two of the most important contributions to twentieth-century physics.

Brownian motion, the small zigzagging movements executed by dust particles immersed in a fluid, had baffled scientists since its discovery by the Scottish botanist Robert Brown in the nineteenth century. Einstein's paper explained it in terms of the millions of collisions that take place between the molecules of the fluid and immersed microscopic particles. This also had important ramifications for the advancement of physics, for it was the first occasion that the atomic theory was shown capable of producing explanations of observed phenomena.

Einstein spent the following years working through his conceptions of space and time, supporting himself and his family in a number of teaching posts in various European universities. After teaching at the University of Berne he moved, in 1909, to the University of Zürich and in 1910 to the German University of Prague, returning to the Swiss Polytechnic in Zürich in 1912. Finally in 1913 he settled in Berlin to become a professor at the University of Berlin and a member of the research institute, the Royal Prussian Academy of Sciences, as well as a director of the prestigious Kaiser Wilhelm Institute. While there, his General Theory of relativity was published. This work, of enormous intellectual and philosophical elegance, developed the concepts he had outlined in the Special Theory into a new interpretation of the Universe. In it, the laws of physics were reduced to those of the geometry of a spacetime continuum. It related gravitation with electromagnetics, two fundamental forces of the Universe that had been hitherto treated as distinct entities. The first proof of his theory came shortly afterwards, when it was observed during an eclipse of the Sun that the light from distant stars is bent as it passes close to the gravitational pull of the Sun.

In his personal attitudes Einstein reflected the growing realization within the scientific community that the scientist by virtue of both his intellect and the nature of his discoveries, has a special responsibility towards society. World War I affected him deeply, and made him very conscious of the outside world and the social injustice and folly that abounds there. Yet he was too much of a genius to ever live completely outside the rarefied atmosphere of his own mind. He was notoriously absent-minded, and anecdotes about him are to be found in abundance in his biographies. One day, for example, a student pointed out to him that he had forgotten to put on any socks. Einstein observed that, since he had not noticed it beforehand, socks must be a quite expendable article of clothing. From that day he is supposed never to have worn socks.

In 1921 he visited America to speak on behalf of the Zionist cause, and in 1922 he was appointed to the Intellectual Cooperation arm of the League of Nations. When it became known in 1939 that German scientists were working on the fission process that would lead to the construction of an atomic bomb, he wrote his now famous letter to President Roosevelt. As a result of this, the Manhattan Project was set up, which resulted in the development of the atomic bomb by the Americans before the Germans. He was deeply involved throughout the war in organizations providing relief for refugees from war-torn Europe. He died at Princeton, New Jersey in 1955.

Perhaps the best epitaph for him is the German proverb he was fond of quoting to express his own convictions about the world: 'God is subtle, but he is not malicious'.

Max Planck

Max Planck is best known for his quantum theory of electromagnetic radiation which he first presented to a meeting of the German Physical Society in December 1900.

Born in Kiel in 1858, he moved with his family to Munich where he attended school and university, later transferring to Berlin University to be taught by the great physicists of the day, Herman Von Helmholtz and Gustav Kirchhoff. Throughout his long life Max Planck was interested in the study of heat, known as thermodynamics. It was in this subject that he presented his doctoral thesis and published papers which led to him becoming a professor at Berlin University on Kirchhoff's death.

Unlike most scientists the breakthrough in scientific thought for which Planck is today remembered came rather late in life when he was 42. This was his discovery of the quantum theory of energy for which he was awarded the Nobel Prize in 1918. Planck was the first to realize that the energy of all electromagnetic waves (including light, heat and radio waves) can exist only in the form of discrete packages, or *quanta*, rather than being continuously distributed in a wave-like form. In this he returned to the 'corpuscular' theory of light which Newton had rejected.

Ironically Planck was both worried and frightened by the theory he proposed. A mild man descended from a long line of lawyers and civil servants, he was cautious in his speech and believed totally in the classical theories of electromagnetic radiation as explained by Maxwell, which were shown by his theory to be inadequate at short wavelengths. Also he was not satisfied with the mathematical formulation of the quantum theory, as it gave the energy of a quantum of radiation as the product of the frequency of the radiation and a small constant. Planck firmly believed that this constant could be removed. *Planck's constant*, as it is now called, is one of the fundamental constants of nature (like the velocity of light in a vacuum) and is vital to the understanding of the nature of atoms themselves and how they absorb and emit radiation.

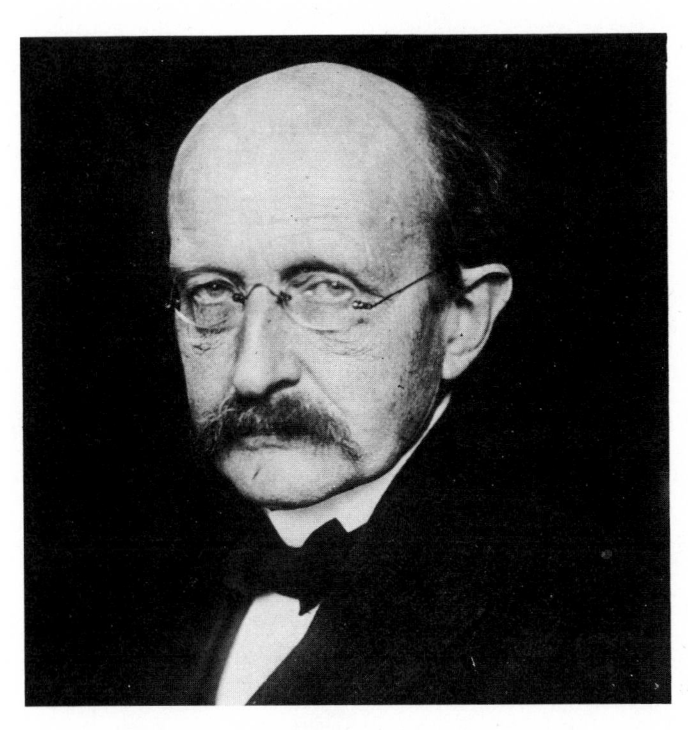

Thus Planck was rather annoyed when his quantum theory was championed by an unknown Swiss clerk in a paper on the theory of relativity. The clerk's name was Albert Einstein. After this bad start a firm friendship grew between the eager young Einstein and Planck, who was, by now, middle-aged. It was said that neighbours often heard them playing chamber music together, Einstein on his violin and Planck at the piano.

Perhaps it was partly this friendship as well as Planck's faith in God which enabled him to withstand the many trials in his life. He continued to teach physics in the then decadent and crumbling Berlin until the age of seventy, even visiting Hitler in his capacity as secretary of the Prussian Academy of Sciences. Sadly, for Planck and for German science, many of his colleagues, including Einstein, had to flee from the Nazis, and Planck's son, Erwin, was executed as an accomplice in the July plot against Hitler.

Max Planck's greatest sorrow was the rift in physics he felt he had caused. Until his death, at the age of almost ninety, he strove to reconcile the classical physics he believed and taught with the modern physics he had founded.

Left: Max Planck, with an extract from a letter of May 1916 to Arnold Sommerfield, dealing with electron transitions in Planck's theories.

INDEX

Pictures supplied by

Dr P S Alpin: 105;
AGA Ltd/Photo: Alan Marshall: A P: 80/81:
Aspect: 24/25
Bell Laboratories: 94
Bell Telephone Laboratories: 59
The Bettman Archive: 114t; 115t
BICC: 79
Ron Boardman: 62
Paul Brierly: 22b; 61b; 108r; 112t
British Petroleum: 34
CEGB: 30; 70; 106
CERN: 98l; 100
Clarendon Laboratory, Oxford University: 58
Colorific: 17r; 24 – Photo: J Pickerell
Daily Telegraph: 46r
Douglas Dickens: 15L
Dupont Company of Canada Ltd: 20tl & 20tr; 21tr & 21 tl
Esso: 26/27; 38
Robert Estell: 44b
Mary Evans Picture Library: 8/9; 18/19
Ferranti: 78
Foto: Hansman Munchen: 110L
The Gas Council: 41t
Hale Observatories: 93b; 114/115
Hart Associates: 23
Michael Holford Library: 93t
Icleandic Photo & Press Service: 42
Dave Kelly: 20b; 21bl
Keystone: 47t
Courtesy of Kodak Ltd: 8
Lurgi: 41b
The Mansell Collection: 113
J A Mitchell Engineering Ltd: 15r
Photo: Michael Newton: 16
Novosti Press Agency: 99bl
Photo Library of Australia: 31t; 43; 46L; 47b; 51; 55t
Photri: 9; 13; 31b; 52t; 55b; 76; 82t; & 82b; 88b; 99r; 104
Picturepoint: 14; 56/57; 89t
Pirelli: 74
Redpoint Ltd/Photo Paul Brierly: 19
Reyrolle Parson: 66
Ronan Picture Library: 11; 92
Royal Danish Embassy: 95t
RTHL: 116t & 116b; 117
Scala: 12
Shell Photo Service: 36; 36
Science Museum: 45b
Snark International: 95b
Staatsbibliothek, Berlin: 117b
Stewards & Lloyds of S Africa Ltd: 44t
STL/ITT: Photo: Paul Brierly: 64
Sunday Times Magazine No. 84: Taken on Kodak Films: 86b; 87
UKAE: 52b; 52/53; 53t & 53b; 54; 65; 91; 98r; 99tl 108L; 109; 112t; 112b
UPI/Popperfoto: 6/7
Roger Viollet: 22t
John Watney: 90r
ZEFA: 32; 33; 35L & 35r; 90L

F